PRAISE FOR
*Return to Treasure Island and
the Search for Captain Kidd*

"A robust and chilling account of Kidd's barbaric exploits."
—*Publishers Weekly*

"If all this sounds like a fascinating adventure story, that's
because it is. Young and old readers alike will find a terrific
pirate tale and fodder for the imagination."
—*Booklist*

"An enjoyable re-creation of Captain William Kidd's last days,
twined with pirate-hunter Clifford's efforts to locate the captain's last great privateering vessel. . . . Clifford still knows how
to wring every drop of romance from his pirate-hunting."
—*Kirkus Reviews*

D1457257

BARRY CLIFFORD is an undersea explorer who discovered and excavated the *Whydah*, the first pirate shipwreck ever authenticated, off the coast of Cape Cod. He established the Expedition Whydah Sea Lab and Learning Center in Provincetown, Massachusetts, where he also owns and operates a pirate museum.

ABOUT THE AUTHORS

PAUL PERRY has cowritten three *New York Times* bestsellers and *Expedition Whydah: The Story of the World's First Excavation of a Pirate Treasure Ship and the Man Who Found Her* with Barry Clifford.

Also by

BARRY CLIFFORD

EXPEDITION WHYDAH
(with Paul Perry)

THE LOST FLEET

Perennial

An Imprint of HarperCollins*Publishers*

RETURN TO
TREASURE ISLAND
AND THE SEARCH FOR
CAPTAIN KIDD

◆

BARRY CLIFFORD

WITH PAUL PERRY

For
Margot
— *B.C.*

For
Reed Bennett Perry
— *P.P.*

A hardcover edition of this book was published in 2003 by William Morrow, an imprint of HarperCollins Publishers.

HarperCollins books may be purchased for educational, business, or sales promotional use. For information please write: Special Markets Department, HarperCollins Publishers Inc., 10 East 53rd Street, New York, NY 10022.

First Perennial edition published 2004.

Designed by Deborah Kerner/Dancing Bears Design

The Library of Congress has catalogued the hardcover edition as follows:

Clifford, Barry.
Return to Treasure Island and the search for Captain Kidd/Barry Clifford
with Paul Perry.—1st ed.
p. cm.
ISBN 0-06-018509-0
1. Adventure Galley (Ship). 2. Shipwrecks—Madagascar—Sainte-Marie-de-Madagascar Island. 3. Kidd, William, d. 1701. 4. Clifford, Barry—Journeys—Madagascar—Sainte-Marie-de-Madagascar Island. I. Perry, Paul, 1950–. II. Title.
G530.A146C55 2003
910'.9'165'23—dc21

2003048774

ISBN 0-06-095982-7 (pbk.)

04 05 06 07 08 ❖/RRD 10 9 8 7 6 5 4 3 2 1

Contents

Contents

PART I

THE FIRST EXPEDITION

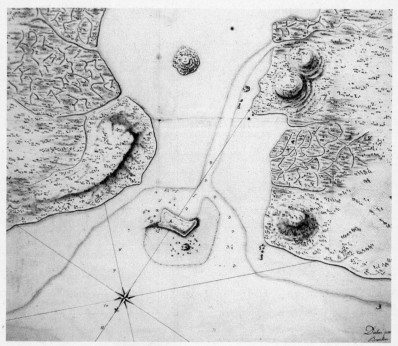

Our trusty and well beloved
Captain William Kidd

—KING WILLIAM III

1

BEING THERE

I FIRST SAW ÎLE SAINTE-MARIE FROM the wing seat of a French turboprop owned by Air Madagascar, an airline with the well-deserved nickname "Air Mad." As the plane began its descent from the west, it dropped through mounds of cumulus clouds before leveling off a few hundred feet over the choppy waters of the Indian Ocean. A few seconds before touchdown, the pilot caught sight of the air sock on the side of the runway—the one near a small herd of zebus being tended by a young boy—and decided the wind was blowing the wrong way. He chose to land from the east instead.

He pulled back on the stick and pushed the throttle forward; the airplane rose, rapidly ascending over the island. Even though the ground flew by fast, it wasn't difficult to see why this had been such prime real estate for the pirates of the East Indies. The runway that we had opted out of seemed to be cut from a lush canopy of foliage, bordered by trees so heavy with fruit they leaned toward the center of the landing zone. We zoomed over an aqua-blue lagoon crisscrossed by graceful wooden pirogues. Farther out into the Indian Ocean, a reef spanned the length of the island for as far as I could see. Surf pounded the reef's ocean side—waves that started in Australia nearly five thousand miles away and rolled unhampered across the third-largest ocean in the world.

"Air Mad" on one of its daily runs to Île Sainte-Marie.
Margot Nicol-Hathaway

The plane banked hard and began a steep descent. Although these aeronautic gyrations were apparently normal on this route, in the United States they would probably qualify as evasive maneuvers.

"*Now* we're flying," said my son, Brandon, a professional skier with a lust for tight turns and steep drops.

The plane righted itself quickly and came in on the short runway. With skillful braking and reverse thrusting, it stopped just before the beach.

"That was thrilling," said Jeff Denholm, a diver, surfer, and triathlete from southern Maine who had lost his right arm in an Alaskan fishing-boat accident. "One of the scariest things I've done this year."

The steward popped open the door, and hot tropical air immediately spilled into the cabin. I took a deep breath and relaxed. The other passengers were unfolding from the tight seats, gathering their carry-on luggage from the overhead compartments, and heading for the open rear door. I sat quietly and let the moment settle in, thinking about the circumstances that brought me here to a place that one historian has called "the only pirate island in human history." *I can't believe it,* I said to myself. *I am actually here. One step closer to finding Captain William Kidd's flagship, a monument to one of history's most misunderstood rogues.*

FOR YEARS I HAD BEEN compiling a file on Kidd. Though his reputation suggests him to be the most notorious and feared pirate of all— "a nondescript animal of the ocean," said a later biography—my research showed that he didn't truly become a pirate until late in his life. To American colonists of the 1690s, Kidd was a pillar of society, a loyal supporter of the king of England and a good seaman who used his skills to steal from enemies of the Crown.

About his early life little is known. He was born 1654 in Dundee, a Scottish seaport. His father was a sea captain who died when Kidd was very young, leaving his family in great poverty. Kidd's ability to navigate and write well indicate that he somehow received a good education. He went on to serve in the Royal Navy, probably as a petty officer. Later he became respected as a privateer, a sea captain who was authorized by his government to rob the ships of the enemy, in this case the French. Kidd was good at what he did. Royal governors in the Caribbean commended him for his fighting abilities, and an English captain familiar with Kidd testified years later at his trial that he "was a mighty man in the West Indies."

Kidd became such a man in New York, too. Rewarded for his bravery at sea, he went on to live in the colonies, where he amassed considerable wealth and respect. His marriage to a wealthy widow gave him higher social standing and added more heft to his bottom line. Kidd became known as one of the movers and shakers of New York City. He owned docks, several town houses in what is now the Wall Street area, a farm in northern Manhattan, cargo ships and businesses. He even helped build Trinity Cathedral next to the site of what would later become the World Trade Center.

Kidd was wealthy, secure, and respected in 1696 when, at the age of forty-one, he agreed to become a privateer for a partnership of businessmen headed by an English lord. Even King William III joined the venture, an act he would later regret. Kidd's goal, as stated in a commission from the king, was to rob French ships and capture pirates who had been plaguing English shipping in the Indian Ocean.

To enable him to carry out his mission, the partnership built the *Adventure Galley*, a hybrid fitted with sails and oars and thirty-four cannons, and the first ship ever built by the British to hunt pirates. Rated at 287 tons, she was light and fast. The oars gave her an extra edge by increasing maneuverability and allowed her to pursue prey on a windless sea.

With a strong ship, a good crew, and financial backing from England's nobles, Kidd seemed to have had everything he needed in order to succeed as a privateer. But appearances were deceiving. After months at sea Kidd realized that he would never make as much money as he had initially thought. There were few French ships to be found, and even if he found them, the agreement he had signed gave most of the profits to his backers.

It wasn't long before Kidd's crew harbored similar thoughts. They began to smolder with anger at the notion of spending months—even years—at sea with little to show in return. Words flew, tempers flared, and Kidd found himself slowly losing control of his own ship. Out of desperation he shifted tactics from "privateering" to "piracy" (although he never considered himself a pirate) and started robbing cargo ships of his own country and its allies.

When word of his crimes reached England, the king was publicly embarrassed. Being involved in a partnership with a pirate was bad enough, but one that robbed ships belonging to the East India Company, an English trading concern that shipped goods to and from the Indian continent, was a disaster. When the mogul emperor of India demanded that the king do something about piracy in the Indian Ocean, King William was forced to put a bounty on Kidd's head. From that point on Kidd was a doomed man.

The *Adventure Galley* became a metaphor for Kidd's life. Rotting from months at sea, its hull being eaten by sea life, the once proud ship was almost beyond repair. But where could he go? He had committed piracy in the Indian Ocean and would risk capture if he put in to any port for repairs. He turned to the only place he could be safe, the island of wanted men, Île Sainte-Marie. It was here, on the only true pirate island in the world, where Kidd's crew would abandon him and force him to burn his dominion, the beloved *Adventure Galley*.

THE STORY OF CAPTAIN KIDD has always fascinated me, and so too has the island where he joined the brotherhood of pirates. Île Sainte-Marie was a dream come true for hundreds of pirates, a safe haven where any man with money could buy clothing, weapons, drugs, alcohol, and women. Captured ships provided a steady stream of goods to the tiny island, and a fortress of cannons guarded the harbor's mouth so that only the pirates and their allies could enter.

Captains harbored on this island hold some of the records for the most booty seized by pirates. In 1695 Henry Avery captured gold and gems estimated by some as being worth more than two hundred million dollars in modern currency from just one Indian warship. The same sources reckon that John Halsey in 1700 took an estimated fifty million dollars from an Indian ship at the mouth of the Red Sea. And in 1721 John Taylor and Oliver La Bouche stole from a Portuguese ship an incalculable hoard of gold and diamonds, one that might well have been worth nearly half a billion dollars today. There are many variables when figuring the modern value of stolen cargo; these estimates are intended only to provide a rough idea of the wealth seized by pirates in this area of the world. The booty taken by Caribbean pirates, such as the infamous Henry Morgan, paled in comparison with that of Sainte-Marie.

Although their thefts had a great impact on the world economy at the time, perhaps their most lasting bequest has been their contribution to literature. It is from their lives that *A General History of Pirates* was written in 1724.

Credited to Captain Charles Johnson (probably a pseudonym of Daniel Defoe), *A General History of Pirates* is believed to have inspired most of the pirate literature we are so familiar with today. Washington Irving drew from it, as did Edgar Allan Poe. But the most famous work to be derived from Île Sainte-Marie and her pirates was *Treasure Island,* by Robert Louis Stevenson.

Stevenson drew so liberally from the work of "Captain Johnson" that, according to historian Jan Rogozinski's pirate history, *Honor Among Thieves,* Stevenson simply relocated Johnson's tall tales from the Indian Ocean to the Caribbean. "In this way, Jim Hawkins and Long John Silver can be counted among the . . . children of the Madagascar pirates."

Although everything Stevenson had read about pirates—including the biography of Captain Kidd and the stories of Indian Ocean marauders—tugged at the back of his brain as he wrote his classic adventure book in 1883, the similarities between Île Sainte-Marie and Treasure Island are the most striking.

The map that he drew for the front of his book includes an island that looks remarkably like Île Sainte-Marie, with the bay at one end in which two smaller islands are placed like gems in a setting.

Treasure Island had a pirate fort and caves, just like Sainte-Marie. At both islands a young boy was cast among the pirates. On Treasure Island it was Jim Hawkins; at Sainte-Marie it was a pirate from Massachusetts named Samuel Perkins.

And then there was the immortal character Long John Silver, who sailed with Edward England before joining Captain Flint. The real-life England and his men spent long stretches between voyages on the beach at Sainte-Marie. In Captain Johnson's profile of Edward England, a ship named the *Cassandra* is captured and the pirates want to hang the captain, a man named Macrae. Before they can do so, one of the pirates, a former shipmate of Macrae's, begins to pace in front of his fellow crewmen, his wooden leg clopping on the deck.

"Shew me the Man that offers to hurt Captain Macrae, for I'll stand by him," he fumes.

It seemed clear to me that Stevenson took elements from the historical Sainte-Marie and transplanted them to the Caribbean, right down to the one-legged man who became known in literature as Long John Silver.

It thrilled me to find these connections between reality and fiction. *Treasure Island* has always been among my favorite books. As a child I reread it frequently, envious of the exciting life being led by Jim Hawkins and intrigued by the colorful and sometimes comic pirates who gave the book such a sinister allure. Reading the book as an adult, I realized that it still held a dream for me, that hope of someday finding a treasure map and following it to a chest of gold.

In assembling research on Captain Kidd I realized that such a map could be put together. By following clues in the historical record I could go to the real Treasure Island and find an archaeological treasure, the sunken flagship of one of the world's most notorious and misunderstood pirates.

To search for the flagship of Captain Kidd at the site of the island that inspired Stevenson's book would be a dream expedition for me. In hopes of making it happen someday, I collected all the information I could on the subject and kept it in a file marked "Captain Kidd/ Treasure Island."

In March 1999 the National Geographic Society presented an exhibit of artifacts from the *Whydah,* the pirate ship I had discovered off Cape Cod and have been excavating ever since. The exhibit, displaying the only documented pirate-shipwreck artifacts in the world, was a resounding success. Thousands of people visited daily, and the *Washington Post* praised its "high-powered presentation that manages to preserve the swash and buckle of piracy."

A few days after the exhibition opened I was sitting in the unassuming office of Michael Quattrone, the general manager of the Discovery Channel, chattering away about the success of the *Whydah* exhibition and my long, strange history with this slave ship turned pirate vessel. Quattrone had called me before I left home (Provincetown, Massachusetts), asking me to come by and see him while I was in town for the opening.

Half a dozen people were eagerly waiting to talk to him, but he was so entranced by the subject of pirates that I felt encouraged to mention the idea of Discovery doing another documentary film on the *Whydah.*

Several television programs had already been made about the noted pirate ship, and another was coming soon from the National Geographic Society.

"No, I think we'll pass on the *Whydah* for now," he said. "We want something else. Something fresh."

But what? We discussed the possibilities, ranging from Blackbeard ("Someone already thinks they have found his ship," said Quattrone) to the "terroresse" of the sea, Anne Bonny ("She left no ship to find," I told Quattrone). The one idea we kept coming back to was Captain William Kidd.

"Everyone knows who Kidd is," said Quattrone. "Or at least they think they do." He wondered aloud as to whether the legend of the most notorious pirate ever was really deserved.

Thinking of my thick file, I assured him that the truth was far more complex than the legend.

Quattrone paused a moment. "Let's do it."

When I returned home Ken Kinkor was the first person I told about my meeting with Quattrone. As the historian for the Expedition Whydah Sea-Lab and Learning Center, a museum founded to display the artifacts of the *Whydah,* Kinkor spends his days and nights culling through historical documents, reading obscure records, and communicating with scholars all over the world in search of any scrap of pirate history. We call Kinkor the "Human Internet," because his brain is so full of information that he sometimes seems to be online.

Quietly tenacious and very restrained, he is the product of his Iowa upbringing. When I told him that Discovery was interested in doing a TV special on Captain Kidd, his mouth curled into a slight smile and he said, "That's an excellent choice. Kidd was a very complex man."

With that, Kinkor turned from whatever he was doing and began to pull together all the information he had on Captain William Kidd—privateer for the king, entrepreneur, husband, stepfather, landowner, and, of course, pirate.

Kidd the man was extremely interesting. But the driving force behind a Discovery Channel television program like this would be the search for an object with a connection to that historical person. That is why I proposed to search for Kidd's second flagship, the *Quedagh Merchant,* which had sunk in the Caribbean.

They agreed, and Ken began the slow process of compiling notes on the whereabouts of the ship. After several days of research he developed a summary report showing the approximate site of the *Quedagh Merchant,* in a river in the Caribbean. "The bad news is that she was burned by Kidd's crew after he left her to go to Boston," said Ken. "The good news is that she is made of teak, so what wasn't burned might still be well preserved."

Teak was good news indeed. English oak, a softer wood from which most English ships were made, would likely have rotted after so many years underwater, but teak is hard like iron and resists being eaten by marine organisms.

With a reasonably accurate location in mind, I felt confident that we could find the ship, and set out to obtain permission from the government of the Caribbean country in whose waters we would search.

My letter to the country's Washington ambassador got me a terse reply: No. They didn't know where the ship was or even if it existed, but they wanted to keep it that way.

"It looks like we're out of luck on Captain Kidd," I said to Ken.

Without the proper permits we wouldn't be able to search for the *Quedagh Merchant*.

"The *Quedagh* was Kidd's *second* flagship," said Kinkor. "Maybe you could look for his first instead."

"The *Adventure Galley*?" I said. Although I had maintained a file on Kidd's first flagship, I was not totally sure of its exact location.

In his research, however, Ken had run across the deposition of Theophilus Turner, one of the men questioned by the British before and during the trial of Captain Kidd. It had long been known that the *Adventure Galley* had sunk in 1698 somewhere near Île Sainte-Marie off the coast of Madagascar. One statement Turner made, however, gave Ken a good idea as to where exactly on Sainte-Marie the final remains of the *Adventure Galley* might rest:

> . . . at the Key [small island] where likewise was a Ship on Ground said to be brought in there by Capt. Kidd & reputed to be taken from the Moors or Turks. . . . There was likewise bottom of the Adventure Galley in which Capt. Kidd sailed from New Yorke and the Ribbs of one other Turkish Shipp taken either by Capt. Hoare or Capt. Glover.

These cryptic words contained crucial information. They established the location of the *Adventure Galley* as being on a small island and in water so shallow that what was left of her was still visible above the surface. Kinkor was sure this remark referred to the "careening spot" on Île Sainte-Marie, an area where ships were pulled up on shore so their hulls could be repaired. Kidd must have assumed that the hull of his leaky ship could be fixed and beached the vessel there.

"I think you could find the *Adventure Galley*," said Ken.

"How sure are you?" I asked.

He mulled over the question for a moment and then gave his estimate. "About ninety-eight percent sure," he said.

He added that he was "dead sure" that we would find pirate ships on Île Sainte-Marie. His reading of the testimony of Kidd's men convinced him that there were several pirate wrecks to be found in the harbor of this tiny island.

"This harbor was a pirate chop shop," said Kinkor. "I think you'll find so many ships that your main problem will be sorting out the *Adventure Galley* from the rest. That's, of course, if there's anything left."

Anything left. Ken had a good point. After more than three hundred years underwater, what would remain of a fully submerged wooden ship? Probably not much, but then, wooden ships weren't all wood. There would be nails, metal rigging, and any nonorganic cargo that might have gone down, including cannons, belt buckles, ceramics—or even gold. We would also find round river rocks, which wooden ships carried as ballast to keep them stable in the ever-rocking seas. Underwater archaeologists call ballast stones the "tombstones of sunken ships" because long after the wood itself has disintegrated they remain behind to mark the spot where a wooden ship has sunk.

I thought about the last shipwrecks I dove on, the French fleet lost at Las Aves Island off the coast of Venezuela. There, nearly a dozen French men-of-war and several pirate ships in their company had piled up on a coral reef, a maritime tragedy in France's history equivalent to our own Pearl Harbor.

Of those enormous wooden warships, little wood was to be seen underwater when I went there in 1998 for the British Broadcasting Corporation. What does remain, however, is a fine collection of iron cannons, anchors, and thousands of other coral-encrusted objects once used by French sailors.

What would be left of the *Adventure Galley*? I didn't know. But when it comes to archaeological finds, pirate ships are the rarest of the rare, and to search for and find *anything* from history's most famous one would be a dream come true and truly a coup.

With that in mind, I decided to have satellite photos made of the area where Ken had pinpointed the wreck—the harbor at Île Sainte-Marie.

Through a satellite-imaging expert I contacted Spot Image, a satellite-reconnaissance firm that specializes in high-altitude photographs of any location in the world. To take these photos, they use satellites formerly available only to government agencies like the CIA. I gave them precise geographic coordinates, and several days later I had detailed photos of the remote island once ruled by pirates.

Satellite photos aren't like the ones you might take of the ground from a tall building or a small airplane. Due to the height they're taken from, they require a different way of looking at details. "Reading" them calls for considerable interpretative skills and, of course, knowledge of what you're looking for. And you have to be careful not to fool yourself into seeing something that isn't there.

We studied the pictures carefully, poring over them for a long time. There were oblong-shaped objects under the water that looked like sunken ships. Several of the ghostly images lined the entrance to the keyhole-shaped harbor. Other possible targets could be spotted outside the harbor and near Pirate Island, which sits in the center of Pirate Bay, or Baie des Forbans.

Back in Maryland at the Discovery Channel offices I couldn't wait to show Michael Quattrone the pictures. He was delighted at what we had found. "This looks like an expedition to me," he said. Almost immediately we began working out the particulars of a voyage of discovery that would take us halfway around the world.

AND NOW HERE WE WERE, streaming across the hot tarmac to collect a load of luggage from the cinder-block building that serves as Île Sainte-Marie's premier-airport waiting room. We climbed aboard a Toyota minivan that took us the five rugged miles to Orchidées Bungalows. We passed grass huts and open cooking fires, coral beaches and tall coconut trees, all the while bouncing down a rough dirt road, the main artery of travel on this primitive Indian Ocean key.

The Orchidées looked like the sort of mom-and-pop hotel that could be found in the fifties along the beaches of Santa Monica. Almost too close to the ocean, it had a yard filled with the wild jungle

ORCHIDÉES BUNGALOWS WAS DUBBED "HOTEL HELL" FOR ITS ABUNDANCE OF INSECTS AND DUBIOUS FOOD. *Margot Nicol-Hathaway*

plants of Madagascar, tamed and trimmed just enough to provide a lush landscape. The brightness of the plants—a rich combination of greens, reds, blues, and yellows—stood in stark contrast to the white bungalows.

When I look back at photos of this place, with its outward appearance of neatness and beauty, it's hard to believe that we almost immediately dubbed it "Hotel Hell."

That name took hold as soon as we ventured into the tiny bungalows and laid sight on the accommodations. Margot Nicol-Hathaway, my fiancée and expedition photographer, called them *"Papillon*-like" in reference to the miserable prison cells of Henri Charrière's autobiography, *Papillon*.

They were painted the drab, flat green that is so popular in many parts of Africa. The mattresses were a landscape of hills and valleys formed by the hundreds of bodies that had used them over the years. Each room contained a tiny desk and a lamp; in many of the rooms, the lamp contained a red lightbulb. When asked why so many of the rooms had red bulbs, the hotel manager offered that the establishment was frequented by prostitutes.

I agreed that the rooms were in dire need of some kind of ambiance. They also needed insecticide. Streams of ants crawled across the floor and up the walls. The gecko lizards lived in peaceful coexistence with the ants, while the quick swipes of their tongues kept the flying-insect population down to a tolerable level. For entertainment we could lie in bed and watch this menagerie of insect life carry on its day-to-day life around us.

Many of the crew became sick from the unsanitary assortment of fish, chicken, and other unknown dishes that issued from the galley. The more cautious members of the expedition took the dietary approach of avoiding fruits and vegetables and eating only foods that were deep-fried. Another daily practice was to take a large swallow of Pepto-Bismol, the stomach antacid that renders bacteria inactive.

What was on the plate wasn't the only horror at the Orchidées. One night I noticed that some of the peanuts we were given as snacks had tiny worms in them. When I asked the bartender what worms were doing inside my peanuts, he looked at me as though he didn't understand the question. "They are competing with you for food," he said. "You must be sure to eat fast."

2

THEFT BY COMMISSION

THE IDEA GRIPPED CAPTAIN WILLIAM Kidd like none ever had. *I shall rob pirates for the king of England!*

He was living in New York at the time. Through his marriage to the wealthy Sarah Bradley Cox Oort, he owned the docks at what is now Seventy-fourth Street and the East River, and knew his way around the black-market commerce that went on unchallenged in full view. It was common to see goods stolen from the moguls of India being bartered to such local merchants as Frederick Philipse, whose handsome manor still stands as a tourist attraction about twenty-five miles up the Hudson River, or the DeLanceys, who became extremely rich as well from such questionable trade. At times Andrew Belcher, Captain Gough, and Peter Faneuil (whose Faneuil Hall is now a centerpiece of Boston's downtown) would go on buying sprees in New York, where the pickings from pirates were so good.

Such "legitimate" businessmen visited the docks with their own ships or horse-drawn wagons and carted away goods stolen on the high seas by men sailing as pirates. At one point it was said by the English Lords of Trade and Plantations that so much illegal exchange was carried on in the light of day in New York that an honest businessman was forced to "slip about" and do his business by night.

Legitimate businessmen, ha! thought Kidd as he watched the illegitimate commerce being carried on under the noses of the local English officials who rarely interfered with the sale of stolen goods. In fact, they were some of the purveyors' best customers. A New York merchant was often a fence for goods brought in by smugglers or pirates. So loose were the laws that the Reverend John Miller, His Majesty's chaplain, considered the local merchants to be little more than thieves, writing in a letter to his bishop, "They are all generally cunning and crafty, but many of them not so just to their words as they should be."

At night Kidd sat in taverns among these merchants and thieves, listening to the salty talk of seas sailed and ships robbed, steeped in the swashbuckling tales of men who sailed with the likes of Thomas Tew and Dirk Shivers, among the most notorious pirate captains of the era. They talked of untold riches in the Indian Ocean, and of its exotic islands, including one known as Saint Mary's. It was there, they said, that a pirate kingdom had been established by Adam Baldridge with the backing of New York merchants. Five miles east of Madagascar and situated near the shipping lanes to and from Arabia and India, she was a perfect pirate stronghold. A ring of coral surrounded the island and protected the inhabitants from attack by pirate hunters, the only access being a narrow harbor with two islands in it.

On one of those islands pirates repaired their ships, pulling them ashore to work on their hulls. The other was a hump of land and coconut trees where the pirates could hide their treasure.

It was a pirate's dream, a true island of treasure complete with compliant native women, balmy weather, and clean water. If it weren't for pernicious tropical diseases like malaria or the risk of having deadly poisons sprinkled into their drinks by a jealous woman, this was a place that some of them could call home.

As Kidd drank in these tales, he wondered if he would ever again dare to have a life so exciting as that enjoyed by these gnarled young men around him. It was 1694 and he was about forty years old, an age

AN EARLY RENDERING OF LIFE ON THE PIRATE KINGDOM OF
ÎLE SAINTE-MARIE. *Artist unknown*

at which most seafaring men were ready to "swallow the anchor" and live the rest of their lives on shore.

Kidd had done that. Moving to New York, he had married a beautiful and wealthy widow, fathered a daughter, acquired money of his own, and become a pillar of society. But now he felt a restless stirring, most likely caused by his friend Robert Livingston.

Both were Scotsmen and had become fast friends in 1692 when Livingston purchased property from Kidd to build a private dock. Lucky for Livingston. In 1694 Kidd sat as the foreman of a grand jury that declined to indict Livingston of trading with the French during wartime, a crime that could have carried a death penalty.

After immigrating to Albany, New York, with his father, the young Livingston became acquainted with many of the Indian tribes and eventually was appointed secretary for Indian Affairs by the king. With 160,000 acres to his name, he was a large landowner in New York and had held a variety of political and military positions, including town clerk, speaker of the Provincial Assembly, and colonel in the provincial forces.

Livingston was a man whose scheme would receive a good hearing in England.

"Here's my idea," Livingston may have said as the two drank port in the living room of Kidd's sumptuous town home near what is now Pearl and Water Streets. "Piracy has become a grave problem for England, one that the king has sworn to deal with. There are so many pirates in the East Indies that the East India Company has petitioned the king to dispatch a man-of-war for the specific purpose of wiping out these rogues."

"So why does he not do it?" Kidd would have asked.

"Our war with France has taken all of our ships," may have been Livingston's reply. "There are no suitable vessels that can be spared for such an expedition."

"So what is your plan?"

"You do it. You become a privateer and rob pirates for the king of England."

THIS WAS NOT THE first time Kidd had done such a thing. The first mention of Kidd in the British historical record is in 1690 and shows him to be a privateer for the king in the West Indies. As commander of the *Blessed William,* a twenty-eight-gun ship, Kidd and Captain Thomas Hewetson engaged six French men-of-war near the island of Saint Martin, where the French ships had cut off an English force about to invade the important sugar-producing island.

Kidd and Hewetson came to the rescue of their countrymen. Facing three-to-one odds, they put up such a fierce fight that the French fled in terror. This kind of bravery was unusual in those days, and the two captains were regarded as heroes.

Kidd led a raiding party ashore to enjoy the spoils of war. He and some of his men carried off two thousand pounds' worth of goods, but when they returned, the *Blessed William* was gone. The men remaining on board had decided to elect a new captain and sail the ship to the East Indies to live a life of piracy. The man elected as the new captain was William Mason. Also aboard was Robert Culliford, a nemesis Kidd would meet again later when he was an accused pirate himself.

Embarrassed and stung by the loss of his ship and its contents (roughly a million dollars by some modern-day estimates), Kidd begged for and was given another ship, the *Antigua,* by the governor of the Leeward Islands. Kidd chased his treacherous crew to no avail.

"Most of the crew were formerly pirates and I presumed liked their old trade better than any that they were likely to have here," wrote Governor Christopher Codrington to the Lords of Trade. "I sent [Kidd] after them, but without success, to the Virgin Islands and to St. Thomas's, where it was most likely that they would have gone to water. The loss of the ship and men, which is serious, could not have befallen us at a worse time."

Kidd chased the "Madagascar men," as the thieves of the *Blessed William* came to be called, all the way to New York, where he discovered that they had transferred to another ship and were now on their way to the Indian Ocean and that mysterious pirate kingdom on the island of Sainte-Marie.

In a decision that benefited his career, Kidd ultimately decided not to pursue the pirate crew further. It was February 1691, and a minor civil war was brewing in New York. Two years earlier Jacob Leisler, a Calvinist merchant, had declared himself governor of the colony and seized Fort James, at the tip of what is now Manhattan. There, with a band of rebels, he was now prepared to hold off the troops that King William sent to back his new governor, Henry Sloughter.

While the platoon of soldiers led by Colonel Richard Ingoldsby attacked the fort, Kidd provided their supply line, running guns and powder to the English garrison.

Sloughter owed a debt to Kidd, and he paid it. As one of his first official acts, the new governor ordered that Kidd be paid one hundred fifty pounds "as suitable reward, for the many good services done to this province." He also presented him some booty taken from the French by members of the *Blessed William*. Ironically, the stolen ship had recently docked in New York Harbor. Kidd argued that some of the booty was rightfully his.

As a bonus, Kidd received a French ship that had been taken by the *Blessed William* as a spoil of war. Kidd immediately sold the ship to merchant Frederick Philipse for five hundred pounds.

Kidd was the toast of the town, and his fortunes kept growing. He started a thriving shipping company, bought land, and quickly became one of the glitterati of society. New York City was unlike anywhere else in the colonies. As a seafaring frontier town of five thousand, populated by an ethnic mix that included French, Germans, Spanish, and Dutch as well as English, it had an international look and feel. Amid the cluster of houses and buildings at the southern tip of Manhattan women dressed in bright colors, used cosmetics, and wore silk stockings, and people spoke freely about politics and religion. It didn't compare with other colonial towns like Boston and Philadelphia, where a more conservative environment prevented free expression.

It was in this tolerant and easygoing place that Kidd found his bride. Her name was Sarah Bradley Cox Oort. The historic record shows that he met Sarah in early 1691 when he bought a home from her at 56 Wall Street.

She was known as a beauty, a woman "being worthy of mention," according to Abraham de Peyster, a public official, in his memoir. She owned one of the finest mansions in town, along with considerable holdings in the city and a summer home on Saw Kill Creek in Harlem. Sarah likely amassed her impressive fortune through marriages, having already been married twice when she and Kidd wed on May 16, 1691, only four days after the death of her second husband.

At the age of thirty-seven, Kidd now appeared to have it all. The family moved into a new town house at 119–121 Pearl Street, where the couple managed their property, which included many docks and warehouses. And Kidd continued to cruise the coast as a sort of colonial coastguardsman, protecting English ships from French privateers.

Life was good for Kidd, and he appeared to be happy. In a letter of recommendation to the king that was written by a colonial-government official, Kidd was described as "a home-loving citizen of New York where he had a wife and a family." Among his friends and acquaintances were many prominent people of the time, including three governors. He and his wife had even made a commitment to buy a pew in Trinity Church, which was in the process of being built.

PERHAPS KIDD'S GREED KNEW no bounds. Or perhaps he was restless, a seafaring man with no legs for land. No one truly knows what it was that drove him to become a privateer, although some think it was his desire to climb the social ladder through wealth and recognition that would ultimately prove to be his downfall.

"So, what do you think of the idea?" asked Livingston again, sitting on the chair near Kidd.

Kidd forced his attentions back into his own sumptuous town house and the fellow Scotsman whose idea would launch Kidd into an infamous doom. "This notion has merit," he said after a while.

"Merit and support," said Livingston. "Piracy has become so bad in the East Indies that the great mogul of India says he will begin to ravage the ships of the East India Company if the king does not do something to stop piracy."

"Excellent idea," said Kidd. "It could not be better. Working for the king to rob pirates. And I get to take French ships as well."

"Yes," said Livingston. "The key, of course, is to never rob a ship that is flying the English flag."

"Of course," said Kidd. "Hands off the ships of the mother country."

Looking out his window at the bustling port of New York, Kidd must have felt a stirring of excitement.

"Livingston, my friend, I am perfect for such a position," he said. "I know every pirate haunt. I know where they lurk. What better thing to do than send a rogue to catch rogues?"

"Indeed," said Livingston. "That is why I came to you with this idea."

3

A Jumbled Graveyard

Since the crew wouldn't arrive until afternoon, I used the free time to discover what a magnificent place this island was. Lord Bellomont, one of Kidd's patrons, must have been thinking of the sheer beauty and natural resources of this island when he said, "The vast riches of the Red Sea and Madagascar are such a lure to seamen that there's almost no withholding them from turning pirates."

I had been here only a day, and already I could feel the freedom and abundance that this island had bestowed on its people for hundreds of years. Margot and I decided to borrow some bicycles from the hotel and set out into the town of Ambodifotatra, the only place on the island with paved roads. The dense jungle to our right at times opened to reveal fields of rice and manioc, which has leaves that taste like spinach and roots that can be pounded into a powdery mash used to make a flat bread. On other fields zebus roamed; these small cousins to the water buffalo are so revered in Madagascar as to be displayed on the nation's currency.

The view to our left was even more extraordinary. Through a narrow forest of coconut trees we caught an occasional glimpse of the Indian Ocean, and beyond that the lush green shoreline of the Madagascar mainland. It was only about five miles away, and when we stopped to take in the view we could see giant trees from the forest that stretched all the way down to the water at a place called Soanierana Ivongo. Further north on the mainland, jutting out like the beak of a phoenix, was Pointe à Larree, part of the land bridge between Sainte-Marie and the mainland that had collapsed thousands of years ago. It was forested with the giant *nato,* a hardwood tree used by natives to build dugout canoes, and from our perspective it looked as though the trees were growing right out of the sea.

As we reached a small bay and began to cross the water on a narrow and rough causeway, I realized that we had arrived. This was Pirate Bay, or Baie des Forbans. Of the bay's two islands, only Îlot Madame was accessible by the causeway, while Pirate Island sat alone in the middle of the bay.

Once we reached Îlot Madame, we dismounted to take a closer look at our surroundings.

From the satellite photos we had examined in Cape Cod I had originally thought that Pirate Island was the careening area. But now, standing on Îlot Madame, I realized that this low and sandy island probably served that purpose. What we could see of Pirate Island seemed too steep and rocky to pull a ship onto. Plus its part of the bay was too shallow and rocky for the deep draft of a large ship.

We began to explore Îlot Madame and found a short stretch of beach that fit perfectly the description of a careening area. It lay by the mouth of the harbor and was sloping and sandy. Behind it stretched a grove of large mango trees. It didn't take much to imagine a ship being unloaded of its cannons, ballast, and heavy goods and then dragged up onto the beach with ropes and pulleys so it could be tipped on its side and the work of hull repair begun.

"Look at this," said Margot when we reached the pier that had been built to the west of what was certainly the careening area. "They've used cannons as bollards on the dock."

I couldn't believe what I was seeing. Emerging every fifty feet or so from the cement the French had poured back in 1946 were old iron

THE CAUSEWAY CROSSING PIRATE BAY, OR BAIE DES FORBANS. IN THE MIDDLE OF THE BAY IS PIRATE ISLAND. *Chad Henning/Discovery Channel*

cannons. Some dated all the way back to the seventeenth century. We found out later that they had been dredged from the bay and stuck into the cement. Rather than import mooring posts, the French must have retrieved cannons from sunken wooden ships, most likely to be found adjacent to the careening area, which, I was now even more certain, must have been right around here.

"Look at this," said Margot again, picking something out of the dirt. It was bone white and looked like a shard of a broken cup. I took it and flipped it over. The outside was covered with a simple blue pattern of flowers.

"It's broken porcelain," I said. We looked for more and found it near what had been the French colonial governor's house. We scraped our feet in the dirt, and more shards emerged. Anywhere we went on Îlot Madame we could find these shards poking up from the ground.

The more we found, the more excited I became. Dr. John de Bry, our project archaeologist, had discovered the logbook of a British naval squadron that visited the island in 1722. Included in the account was "wrecks of several ships which the Pyrates had demolished," as well as "China ware, rich Drugs, and all sorts of spices lying in great heaps" upon the beach.

Although time and sea life had swallowed the ships that the British were talking about, and the "rich Drugs" had been dissolved by the elements, porcelain is forever. Such a large amount of it told me that we were standing at the very spot described in that logbook.

I felt a chill of excitement as I looked over the water that covered what was certainly the wreck site of the *Adventure Galley*. If everything I had studied was true, then there were several shipwrecks out there, literally within a stone's throw. We were about to dive on not just one possible ship, but several. Exciting, yes, but not necessarily the best circumstances for identifying a very specific ship, Kidd's *Adventure Galley*. Although the ship was distinct in certain ways—she was English-built and had a smaller ship with her that sank at the same time, for example—a jumbled graveyard of ships and ship parts would make the *Adventure Galley* more difficult to locate.

It is rare in underwater archaeology to have too much to look at, so I should have been delighted. But I wasn't. The Discovery Channel had given us only a few days to locate the target, and the thought of sorting through several ships to find it was a daunting one.

As we looked out at the harbor, I shared my concerns with Margot. She smiled and said, "The more ships, the more artifacts."

It suddenly didn't seem like such a bad situation.

THE LURE OF HISTORY is every bit as strong as the lure of money, sex, or power, and it had driven me to search for shipwrecks for many years. Carl Jung said, "History is not contained in thick books, but lives in our very blood."

The expedition members getting off the plane that afternoon

would have agreed with the sage psychologist as well. I had worked for several years with all but one of these people, and despite the hard and difficult conditions, they all kept coming back.

Wes Spiegel was the first off the plane. A quiet, rawboned Cape Codder, Spiegel immediately pronounced the island "a pirate's Miami." I have employed him as diver, deckhand, crew chief, even cook in the past, and while he's generally enthusiastic in any assignment, he loves the more dangerous challenges.

Behind him came Eric Scharmer, a gaunt and intense Boston-based filmmaker who worked for several seasons on the *Whydah*. He had his own film and video business and continued to test the limits of his profession by making movies under very dangerous conditions. Scharmer has jumped from helicopters in the Swiss Alps to make ski films and dove hundreds of feet underwater in the South Seas to film the skeleton of a whale that had become trapped in a cave.

Next off was Bob Paine, a marine surveyor from Scituate, Massachusetts. A diver and licensed sea captain, Paine and I had met a year earlier in Provincetown. He had volunteered to help search for *Whydah* artifacts, and I quickly took him up on the offer. Like all of us, he was drawn to the rush of finding something so lost that no one knew whether it even existed anymore. When I offered him the chance to come to Madagascar, Paine dropped everything and packed a bag.

A few locals got off, and then out popped Charlie Burnham. His height, about six feet two, and the tangled mass of Medusa-like hair atop his head make him stand out in any crowd. He tried to look calm as he crossed the tarmac, but Charlie is a nervous flier and the maneuvering that had just taken place clearly unsettled him, evidenced by his damp forehead. A computer and electronics expert with a degree from Yale, Charlie never rests on his laurels. In addition to working with me on several shipwreck finds, he has worked on the *Titanic* site, diving nine times to the deeply submerged luxury liner on a French minisub.

Last off the plane were John de Bry, the project archaeologist, and Paul Perry, who would be covering the expedition for the Discovery

Channel website. I had worked with Perry before when we had coauthored *Expedition Whydah,* the book about the Cape Cod pirate ship, but de Bry was a new face for me, though his presence came with a history.

In 1998 I'd led a British film team and a small group of explorers to the remains of a fleet of French and pirate ships that crashed onto a reef near Las Aves Island in Venezuela, a story described in *The Lost Fleet.* Our expedition attracted considerable media attention, largely because we filmed so many sunken artifacts from these magnificent men-of-war.

When I returned to Cape Cod, I received a letter from de Bry accusing me of "intellectual theft," claiming that we were working from a copy of a map that was his.

What de Bry didn't realize was that his "rare map" had been published in a French textbook, of which we had a copy. When I explained the situation to him, he responded with grace and good humor. The two of us became friends, and I decided to involve him in the quest for Captain Kidd's flagship.

In addition to being a historical archaeologist, an expert paleographer, and a former U.S. Army Airborne officer, de Bry also spoke fluent French. At age fourteen he lured undersea explorer Jacques Cousteau to his parents' home on the Mediterranean coast to help him excavate the remains of an ancient Roman ship.

I was delighted that de Bry had agreed to join the expedition. His primary task would be to record the findings of the team and direct other archaeological tasks such as artifact identification and site evaluation. In addition, because of his language skills, diplomacy, and toughness I had a hunch that he would get us out of some tight spots in the coming months of our expedition. That proved to be true.

The last man on the crew was Gilles Gautier, the puckish French guide I had hired over the Internet. Gilles ran a company in the capital city called Les Lézards de Tana, specializing in Madagascar adventure tours. There were few places he hadn't been in this mysterious country, and he would be the best person to guide us on an island with few decent roads and plenty of rough terrain.

That night we struggled through our first meal at the hands of the Orchidées' chef. In honor of our arrival, he whipped up his best dishes. In the bowls that arrived from the barely serviceable kitchen came mounds of rice that were specked throughout with dirt and insects and a serving plate heaped with an extremely bony fish that none of us could identify. Adding to the problem was a thick green sauce that covered the catch like some culinary camouflage.

"Oh, boy, Malagasy bonefish," said Bob Paine, daring the first taste.

The rest of us dug in slowly and then ground to a halt. The green sauce had a spoiled taste to it, and the fish was so bony as to be impenetrable. Some continued to down the rice while others ordered a bottle of Three Horse Beer and called it an evening. Dinner was over almost before it started.

After that it was early to bed for most of us. Jet lag and the Lariam— an antimalaria medication that I made certain everyone was taking— had worn us out. I told the crew to set their alarms for seven in the morning and headed for our bungalow. Passing the kitchen, I decided to peek in and examine the facilities. I turned on the light and then stepped back in shock. The dishes from the evening meal were still piled high with uneaten food and stacked in the sink. Cockroaches ignored me, devouring the food we'd rejected.

"Why haven't these dishes been washed?" I asked Gregory, the hotel's steadfast security guard, a man who worked virtually twenty-four hours a day and who later became the security guard for our expedition.

"We leave them there to feed our relatives," he said. "They roam the island as ghosts all night, and we leave them our uneaten food. Here, Mr. Barry, even the dead must eat."

4

PRIVATEER, INC.

 IN THE 1690s NEW YORK CITY WAS A RAW colonial port with five thousand citizens, a harbor full of trading vessels, and sailors from all over the world. While much of the trade that went on was legitimate, a substantial amount was the highly profitable, illegitimate trade that English law turned its back on. Merchants commonly made 100 to 400 percent profit on booty purchased from pirates. Piracy was practiced so openly in New York that some politicians suggested making the island of Madagascar a colony so that trade with pirates would be legal.

Under the watchful eye of Governor Benjamin Fletcher, pirates went in and out of New York Harbor by the hundreds. Fletcher's intent wasn't enforcing the law, it was collecting as much graft as possible, and he pursued it openly from every questionable ship that came into port. It was widely known that the good governor could be bought for one hundred Spanish pieces of eight, and it was gladly paid as part of the cost of doing business in the colonies.

With the governor as their guide, pirate businesses sprang up everywhere in the colonies. The tip of Long Island became a rendezvous point for pirates, as did Block Island, and both areas blossomed with pubs and other businesses to take advantage of the buccaneers. A Major Selleck had a warehouse in Stamford, Connecticut, that was famous for storing pirate goods. Boston judge Samuel Sewall operated an unlicensed mint in Boston that was used openly by pirates to melt down stolen gold and silver.

Piracy became so open that the colonies' top preacher, Cotton Mather, spoke out loudly and often on its scandals, never mentioning, of course, that he had a lucrative pirate-related business of his own. He wrote up the confessions of condemned pirates and published them in pamphlets that he sold at their hangings.

In a way, the colonies were what the Wild West would become one hundred and fifty years later. Dozens of desperados were recognized and revered, especially in the streets of New York. Thomas Tew, one of the wealthiest of all pirates, regularly dined at Governor Fletcher's mansion, where he was "received and caressed," according to a report given to the Lords of Trade.

It was the perfect environment for Kidd and Livingston to start a privateering business.

IN DECEMBER 1695, Colonel Robert Livingston left New York, bound for England on the *Charity*. It was a voyage that the tough Scotsman would not soon forget.

Besieged in a violent Atlantic storm, the ship lost her rudder in ravaging waves. For four months she drifted aimlessly, caught in the flow of trade winds and currents that took her first one way and then the other. By the time she finally struck shore in Portugal the crew and passengers were nearly delirious from fear and lack of provisions.

Livingston regained his strength and wits after several weeks in Portugal and then continued his voyage on another ship. This trip was uneventful, and by July, the politician and government official from the new colonies set his foot on a solid English dock in London.

Livingston immediately presented his privateering idea to Richard Coote, the first earl of Bellomont. King William III had just appointed him governor of New England in an attempt to get rid of the crooked Fletcher and to strengthen royal control of the colonies. William had recently seized the throne from James II as a result of the Glorious Revolution of 1688 and was very concerned about maintaining loyalty to his rule in the face of potentially dangerous opposition.

Richard Coote was Lord Bellomont's given name. He had been born in 1653, with not a silver but a pewter spoon in his mouth. His father, also a Richard, was a colonel in the English army who helped restore Charles II to the throne. Out of gratitude, the king granted him the title of first earl of Bellomont. With the title came a large estate in Kerry, Ireland, and virtually no money. Still, the senior Coote held a variety of midlevel government jobs. He was a member of the Committee of Privileges and Grievances, was appointed the commissioner in Athlone for examining Irish delinquents, and was a major in the Regiment of Horse. It was here that the major gained the dubious reputation of being the soldier who burned down the pilgrimage statue of Our Lady at Trim in Tipperary, where hundreds of Catholics went yearly for the supposed healing powers of the icon.

Like his father, young Richard had his wild moments, too. In 1677 he discovered that he was not the only one vying for the hand of a fair maiden named Mary Rawdon. The other suitor, Colonel Alexander McDonnell, challenged Coote to a duel. Coote accepted and was seriously wounded in the competition. The only saving grace for Coote was that he killed McDonnell instantly with a well-placed shot.

Coote did not marry Rawdon. Instead, in 1680, he wed Catharine Nanfan, the daughter and heiress of Bridges Nanfan. It was through her that Birtsmorton Court in Berrow came into the Coote family.

With little but his title to go on, Lord Bellomont became a member of Parliament as a Whig. He went on to devote his service to the royal family, first as an army officer for Prince William and then as treasurer to Queen Mary.

Although he was always surrounded by money, Lord Bellomont never seemed to have any. His land in Ireland was, for the most part, not arable. And although luck came his way, it usually just passed him by. In 1694, for instance, King William granted Bellomont title to more than seventy-seven thousand acres of land in Ireland. The land was given to a grateful Bellomont because of his "necessitous" condition and services he had performed for William prior to William's seizure of the throne. But when Parliament heard about the land transfer, it became so angered that the king was forced to rescind his gift.

Bellomont arrived in the colonies in despair to discover that he would receive no salary. He petitioned the Lords Justices, requesting twenty-five hundred pounds in wages. The Lords of the Treasury agreed to pay him twelve hundred but only once and in lottery tickets at that. The skimpy offer depressed Bellomont, who wrote of his situation to his business manager, Sir John Stanley: "I shall have to return home as rich as I came abroad, which gives me many a Melancholy reflection."

The implications of Governor Fletcher's complicity with pirates were aggravated by pleas from the East India Company that it was literally being ravaged by pirates. They, along with the great mogul of India, demanded that English warships sail to the Indian Ocean and put an end to piracy. The great mogul had suffered greatly at the hands of pirates. Not only had the frequent raids ravaged his personal fortune, they had devastated those of his subjects, many of whom had been personally subjected to pirate atrocities on the high seas. The great mogul blamed the English for all acts of piracy on the Indian Ocean, and at one point he became so angry that he arrested English managers from the East India Company, holding them personally responsible. Now he threatened King William: either he stop piracy, or one of his most important trade routes would be cut off.

Bellomont's orders from the king included explicit instructions to stop the growth of piracy springing from the American coast. Bellomont promised that he would give it his best effort, but in reality he had no real idea how to accomplish this. England was at war, again, with the French, and there were few ships and soldiers available to police the colonies, let alone the Indian Ocean.

Thus, Livingston's proposal couldn't have been better timed. With

Kidd a privateer, the Indian Ocean would be patrolled, pirates would be captured, and the great mogul would be pacified.

Thinking along similar lines, the king approved the privateer venture, and Livingston began to pull together a number of investors who would purchase a ship for Kidd and fund his battle against piracy (and the French) while making a little money for themselves in the process. The king reserved a tenth share in whatever booty was captured. Eight other partners joined in the venture, including the lord high chancellor, the first Lord of the Admiralty, and two of the king's principal secretaries of state. Between them they raised six thousand pounds, enough to build Kidd a new ship for a foray to the Indian Ocean.

It seemed perfect, but Kidd didn't like the terms. According to their plan, Kidd would agree to sail the Indian Ocean and defeat the pirates, taking as many French ships as he could in the process. He would then pack up their booty and return it to Lord Bellomont in New York. In exchange for a small investment and management of the partnership, Bellomont and his partners would receive 60 percent of the profits derived from the booty. The remainder would be divided between Kidd and his crew. If the booty totaled more than one hundred thousand pounds, Kidd would be allowed to keep the ship they were going to purchase for him. If not, the ship would be returned.

It was a unique agreement if only for the fact that a partnership of government officials was going to be privately enriched through the theft of other people's stolen property.

Kidd, who was in England now, consulted his friend Captain Hewetson, who advised against the deal, saying it favored only the investors. Lord Bellomont heard what Hewetson advised and brought considerable pressure to bear on Kidd. Finally Kidd relented and agreed to become a privateer, as much for the wealth as for the improvement of his social status in the colonies.

On October 10, 1695, Bellomont had articles of agreement drawn up that bound him, Livingston, and Kidd together in a legal contract. This agreement would later come back to haunt Bellomont and everyone else involved. At the time, however, he could not have been a happier man.

Bellomont and the partnership he headed agreed to obtain the necessary commissions from King William and pay for a ship for Kidd to command as well as "furniture and victualling." In return, Kidd agreed to "procure and take with him" one hundred sailors and "to make what reasonable and convenient speed he can, to sett out to Sea with the said Ship, and to sail to such parts and places where he may meet with the said Pirates." Furthermore, he was expected to "take from them their Goods, Merchandizes and Treasure" and take it to Boston without stopping at any port or "without breaking Bulk, or diminishing any part of what he shall so take or obtain, on any pretence whatsoever."

With the agreement signed, the Castle's Yard shipbuilders at Deptford on the River Thames began to build a vessel that would play host to Kidd's voyage. She was named the *Adventure Galley*.

5

TREASURES TO EXPLORE

 IN THE MORNING WE ALL GATHERED around the Orchidées' rickety dining room table and examined a map of the harbor. At its widest point the harbor's mouth was about a half mile. Then it narrowed like a funnel toward the bay. Later we would use high-tech methods to map the entire harbor floor, but today we were going to focus on an area about the size of a football field directly in front of the careening spot. It was there, our research led us to believe, that the remains of the *Adventure Galley* could be found.

"Luckily they didn't careen the *Adventure Galley*," said John de Bry, offering the crew a short course in eighteenth-century ship repair. "Ships that came to be careened were almost always completely unloaded before they were pulled up on the beach. And I mean completely. Even the ballast stones in the bottom of the hull are deposited on shore. Cannons, cargo, food, utensils, everything is brought out to make the ship lighter. That didn't happen with this ship. The only thing these sailors wanted to unload was their share of the booty, according to the historical record. When they did that, they said good-bye to Kidd and just left the ship there."

Since it hadn't been unloaded in preparation for repair, de Bry said we could expect to find large amounts of "cultural material,"

Archaeologist John de Bry *(left)* and technician Charlie Burnham plan the first dive with Barry Clifford. *Margot Nicol-Hathaway*

items like china and other nonperishable goods—stolen most likely—as well as metal fittings, like nails and hinges, and cannons. Some pieces of the wooden hull might be found, too. But most certainly whatever we found would be in or around a large mound of ballast stones. Those large piles of ballast stone is what we would be looking for that morning.

Most of the crew climbed into the passenger van for the ride to the wreck site. Others kept an eye on our gear in the bed of the equipment truck. Charlie Burnham had rented a motorcycle so he could travel without having to rely on the native drivers.

The Discovery Channel's camera crew had arrived the previous evening in a Cessna large enough to accommodate all of their equipment. Led by producer David Conover, a friendly giant from Maine who owns Compass Light productions, the three-man crew hired a bat-

tered French sedan driven by a kid we called "Turbo." His leaden foot
and disregard for potholes made him first to almost any destination.

We caravanned to Îlot Madame and assembled on the dock. We were
met there by Maximo Felice, a swarthy diver from Rome, who ran a dive
shop in town and had agreed to make his dive boat available to us for
the duration of our expedition. Now we loaded it with the hundreds of
pounds of scuba tanks, wet suits, and other dive equipment that we had
wrestled onto the airplane in Boston. On the boat Wes Spiegel and I discussed which areas we would examine, while Eric Scharmer prepared
his underwater video camera to record the initial dive.

Our plan was to swim search lines, a process of swimming rows to completely cover an area similar to the way a lawnmower would cover a front
yard. Some divers liken this search method to looking for specific targets
from an airplane. It is tough work and requires constant attention to
detail. Because directions can be uncertain underwater, you have to watch
a compass to make certain where you are heading. You also have to stay
deep enough to keep the sea floor visible while you search for telltale signs
of a wreck.

We put on our wet suits and tanks and sat on the railing of the boat as
I pointed to landmarks that outlined the search area. By using such landmarks, the divers could rise to the surface if they became disoriented and
immediately know where they were. Then, with the remainder of the
crew watching from shore, we flipped off the boat and into the water.
With no fanfare, the search for the *Adventure Galley* had begun.

WATCHING SOMEONE SWIM SEARCH lines is no more exciting than
reading about it. That is especially true if you are an archaeologist like
John de Bry and have swum them dozens of times yourself.

Forced to watch from the shore because a slight bout with malaria
was keeping him out of the water, de Bry was antsy. For about an hour
he stood on the shore with Paul Perry, pacing back and forth along the
dock as our bubbles rose in the harbor.

"Let's go scare up some action," he finally said.

De Bry and Perry strapped on daypacks and, with no particular destination in mind, began walking down the island's only decent road, wracked with potholes.

As they passed one tiny hut that was emanating a reggae version of Sonny and Cher's "I Got You, Babe," a thin man with a broad smile stepped out from behind the lace tablecloth that was covering the doorway and stopped them underneath a coconut tree.

He struck up a conversation in French with de Bry, and when he discovered the purpose of our expedition, he became very excited.

He told de Bry that he was the descendant of a pirate, like so many of the island natives. When de Bry asked him which pirate, the man shrugged. "I don't know his name, and most of us on this island who are descendants of the pirates don't know their names, but we know from our families that we have their blood in us," he said.

That may have been true, de Bry told Perry. At the time that Captain Kidd was here, more than a thousand pirates were reported to be living on this tiny island. Their relationship with the natives varied. Many of the local women married or cohabited with the pirates, living in the same-style grass huts with platform floors that are still popular on the island. The locals often traded happily with the pirates. The islanders' safety from attack by other tribes was assured by the presence of the well-armed foreigners, and the pirates had a useful base of operations from which to prowl the Indian Ocean.

On the whole, the pirates and the islanders had a mutually beneficial relationship. But even so, relations were at times dicey. In 1697, before Kidd arrived on the island, many pirates including Captains John Hoar and Robert Glover were killed when they tried to kidnap some of the islanders to sell as slaves. Good relations were replaced by mistrust and sometimes an outright fear of dealing with the pirates.

"Even now, many people are afraid of the bay and even Pirate Island," said the man, who introduced himself as André Mabily. "The pirates lived there, and there is some feeling that many of their spirits still roam around."

ISLAND GUIDE ANDRÉ MABILY AND JOHN DE BRY EXAMINE A PIRATE
TOMBSTONE. *Paul Perry*

Many here have seen ghosts of pirates in recent years, insisted
Mabily. And some have even died after seeing them. One such phan-
tom, "the red ghost," had been seen walking down the road to the
pirate cemetery by several people who later died, claimed Mabily.

Upon hearing this—especially the part about a pirate cemetery—de
Bry and Perry begged Mabily to take them to see it.

They left the main road and walked for a half hour through thick
jungle that gave way periodically to the front yards of tiny grass huts and
small plots of tilled land. A sign marking the entrance to the Cimetière
des Pirates was on the ground next to two crumbling gravestones. Past
the sign were fifty or more graves of pirates and sailors from England,
France, and other countries. Age and desecration had made it difficult
to read the epitaphs on most of the headstones. The cemetery had the
look of a place that was not visited frequently or cared for often. Several

of the thick stone slabs that covered the human remains had been pried up by treasure hunters searching for any gold and silver that may have been buried with the pirates underneath.

"What happened here?" Perry asked Mabily.

"That is the work of Bi Bq [pronounced Bee Beek]," he said, shaking his head sadly. "But he is no longer someone to worry about."

He explained that Bi Bq was a wealthy man from Réunion Island who had come here several years ago with a crew to search the island for treasure. He began tearing up the cemetery, breaking many of the stones and digging up bodies. It was a horrible mess.

"What was he looking for?" asked Perry.

"Pirate treasure," answered Mabily. "He was a very bad man who thought he could come here and dig anywhere he wanted."

"What happened to him?" asked de Bry.

"The police forced him to stop, and then they asked him to leave. And then . . ." Mabily's voice trailed off, and he grinned as he chewed a piece of grass.

"And then what?" asked Perry.

"And then he was killed, murdered when he got back to Réunion Island," said Mabily. "He was a rich and famous man on his island, but the murder has never been solved."

"So did someone from this island sneak over and kill him?"

"What Bi Bq did was *fady* [taboo]. It is the kind of death that can happen to someone who commits such an act on the dead or the property of the dead. Many people here practice voodoo. People like Bi Bq need to be careful no matter how rich and powerful they think they are."

Mabily gave the two expedition members a tour of the remainder of the graves. The south end of the graveyard was wooded. At the north end the trees fell away and offered a panoramic view of Pirate Bay. From this lookout, Perry and de Bry had a clear view of the crumpled piece of real estate in the middle of the bay known as Île aux Forbans, or Pirate Island.

De Bry mentioned to Mabily that in many of the historical tales about Sainte-Marie, Pirate Island was always rumored to have caves in which the pirates hid themselves or their treasure. As he spoke, Mabily nodded in agreement. "Oh, yes, they are there," he said enthusiastically. "I have been into them myself. I went deep into one, and when it got cold I became frightened and came back out. But they are there. They go almost straight down into the earth."

It must have been the look of disbelief on the faces of de Bry and Perry that caused Mabily to add: "You don't believe me? I will take you to the island and show you that what I am telling you is true."

6

THE ROGUE'S ROGUE

THE *ADVENTURE GALLEY* WAS A PERFECT ship for privateering. Rated at 287 tons measured cargo volume, it was large enough to accommodate a crew of one hundred fifty men, yet small enough to be dragged ashore and careened, an important feature in warm waters where barnacles built up quickly and teredo worms ate the wood like it was butter.

She carried thirty-four guns, each cannon firing a three- to four-pound ball. She was maneuverable and had the added advantage of being a true galley, which meant that she was equipped with forty-six oars that could propel her at about three miles per hour when there was no wind to fill her sails. Using the oars, she could bring her guns to bear on a target very quickly or pursue an enemy on a windless sea.

Before taking charge of his ship, Kidd was given two commissions by King William. One was a privateering commission, a boilerplate authorization that permitted him to attack and bring in enemy ships, specifically those of France. The other was an extraordinary one that empowered Kidd to capture pirates. Like the privateering commission, it was issued under the Great Seal of England, which showed the great import King William attached to this project.

Citing buccaneers who, "against the Law of Nations, commit many and great piracies, robberies and depredations on the seas," the king's

special commission granted Kidd the power to "bring the said Pirates, freebooters, and sea rovers to justice."

Although the directive gave Kidd the authority to "apprehend and seize" those he suspected of being brigands, it also issued a stern warning to Kidd to avoid any acts of piracy of his own: "We do hereby strictly charge and command you, as you will answer the contrary to your peril, that you do not, in any manner, offend or molest our friends or allies, their ships or subjects, by colour or pretence of these presents, or the authority thereby granted."

WITH COMMISSIONS IN HAND, Kidd made handwritten recruiting posters for anyone who was interested in going to sea as a privateer.

With the brashness that he was known to display, he called for able-bodied seamen, announcing loudly in public places that he, Captain William Kidd, was being backed by the full force of none other than the king of England. "We are sailors for the king and for England," he undoubtedly said to the curious. "I have the authority to capture pirates and take their prizes, by force if necessary. We will become heroes, and we will become rich. Join me."

Although the promise of wealth and fame surely sounded good, the real terms of employment were much shakier. Crew members were being offered a "no purchase, no pay" arrangement, which meant that they would be paid only by shares of booty, if any was actually taken. If there were no pirates or enemy prizes to be had, the crewmen would walk away empty-handed at the end of the voyage and the ship would have to be returned to her owners. If the crew of the *Adventure Galley* took considerable booty, they would end up splitting about 25 percent of the proceeds. At its best this was a poor agreement, especially since most other "no purchase, no pay" arrangements had the crew splitting 50 percent.

The agreement eventually included the standard form of workmen's compensation (one hundred pieces of eight for the loss of a finger or toe) as well as an inducement of one hundred pieces of eight for "the man who shall first see a sail, if she prove to be a prize." On the other hand, cowardice or being "drunk in time of engagement" would result in the loss of a seaman's share.

Nonetheless, Kidd's offer had plenty of takers, since a royally commissioned privateer was an attractive alternative to naval service in a time of war. He reported to Bellomont that he had been or would be able to select a choice crew from the many seamen who applied. By the beginning of February 1696, more than one hundred men were ready to sail.

Lord Bellomont watched closely as Kidd raised his crew, for the aging privateer was his meal ticket. If Kidd could capture just a small portion of the booty being taken by Indian Ocean pirates, then Bellomont would make a considerable amount of money. If not?

Bellomont had that angle covered, too. The agreement stated that if Kidd didn't return with at least one hundred thousand pounds in booty by March 25, 1697, he and Livingston were obligated to reimburse Bellomont's investment. If Kidd returned empty-handed, Bellomont had taken out a form of insurance that could net him as much as thirty thousand pounds (fifteen million dollars by today's standards). For Bellomont and the other partners, it was a "win-win" situation.

Still, Lord Bellomont was concerned about Kidd's past. He had been a privateer before entering colonial high society, and some of the privateers he had associated with in his Caribbean days were now Indian Ocean pirates. One of them, Robert Culliford, had even stolen Kidd's ship and used it as a pirate vessel to raid East India Company ships. Was there a chance that Kidd could follow in their footsteps? A chance that this arrogant self-styled captain might feel very much at home in the playground of pirates?

What is that old saying? Lord Bellomont pondered. *Lie down with dogs, get up with fleas.* In the end, however, Bellomont certainly thought this project was worth the risk.

AND SO IT WAS. With the *Adventure Galley* fully stocked with provisions and staffed with a crew ready to embark from London's docks, Lord Bellomont issued a directive to sail on February 25, 1696.

Captain William Kidd,

You being now ready to sail, I do hereby desire and direct you, that you and your Men do serve God in the best Manner you can: That you keep good Order, and good Government, in your Ship: That you make the best of your Way to the Place and Station where you are to put the Powers you have in Execution: And, having effected the same, You are, according to Agreement, to sail directly to Boston in New England, there to deliver unto me the Whole of what Prizes, Treasure, Merchandize, and other Things, you shall have taken by virtue of the Powers and Authorities granted you: But if, after the Success of your Design, you shall fall in with any English Fleet bound for England, having good convoy, you are, in such case to keep them company, and bring all Your Prizes to London notwithstanding any Covenant to the contrary in our Articles of Agreement. Pray fail not to give Advice, by all Opportunities, how the Galley proves; how your Men stand, what Progress you make; and in general, of all remarkable Passages in your Voyage, to the time of your Writing. Direct your Letters to Mr. Edmund Harrison. I pray God grant you good Success, and send us a good Meeting again.

Bellomont.

Kidd had barely begun his journey before trouble found them. At the mouth of the Thames River, he overtook a royal yacht, the *Duchess,* and refused to strike his colors. Striking colors was a show of subservience and courtesy that military ships insisted upon. Kidd felt that he should not have to pay respects to a ship of the line because he had been granted royal authority, which in his mind made him equal in rank to any of the king's commanders.

As the *Adventure Galley* pulled ahead of the *Duchess,* the latter's captain shouted a second order to strike the colors. When Kidd refused, he fired a shot at the *Adventure Galley* and missed. Kidd's colors remained high.

As the *Adventure Galley* gained on the *Duchess,* Kidd ordered his men in the top masts to drop their pants and moon the yacht. A short while later, Kidd overtook another naval vessel and again would not lower his colors. This time he was overtaken and stopped. As punishment, the ship's commander, Captain Stuart, took most of Kidd's men and pressed them into the service of the Royal Navy.

Suddenly Kidd had fewer than half the men he had started with and his mission was in jeopardy. Outraged, he brought his ship into the town of Sheerness and lodged a complaint with Admiral Edward Russell, another of his financial backers. It took nearly a month for the matter to be settled, and while a crew was returned to Kidd, it wasn't the handpicked one he had assembled in London. These were "lower types," he would later say, not as experienced and professional as his original choices.

The mooning incident cost Kidd time and money. It must also have caused Bellomont and the other backers to wonder just how arrogant their Captain William Kidd was. Here he had provoked the Royal Navy on its own turf. What would he try to do when he was a long way from home?

7

A LIKELY SUSPECT

IT WAS NEAR THE END OF THE SECOND day of diving, and the dive crew was worn out.

We had swum back and forth for several hours now, checking out possible sites that had been clearly visible on the satellite photos. There were so many likely targets on these photos and on old charts that John de Bry had found in French archives that I remember saying that it was almost like having Captain Kidd's address.

I was wrong.

The "wreck sites" we'd found on the first day were not panning out. One of them turned out to be the remnants of a dock. Another target turned out to be a rock. A third was just plain nothing. That night we'd gone over the charts and satellite photos again, but to little avail. We went to bed shortly after dinner so we could get an early start the next day. Instead of sleeping, I sat up and pored over the maps again and again. I knew the *Adventure Galley* was right there underneath our noses.

In the morning I was nervous and perplexed. The harbor seemed so devoid of shipwrecks that I was beginning to wonder if the research we had drawn on to get us here was completely off base.

I surfaced next to Wes Spiegel. "What have you found?" I asked.

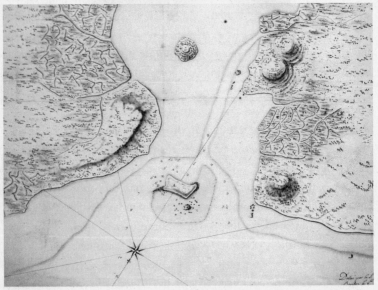

A MODERN-DAY SATELLITE IMAGE OF ÎLE SAINTE-MARIE SHOWS THE ENTIRE ISLAND, WHILE A SEVENTEENTH-CENTURY MAP OF PIRATE BAY SHOWS ÎLOT MADAME AT THE MOUTH OF THE BAY AND PIRATE ISLAND IN THE CENTER.
John de Bry

"Nothing serious," he said. Spiegel is a veteran shipwreck diver, and I could tell from the look on his face that he meant, quite literally, nothing.

"We still have plenty of area to cover," I said optimistically.

"Then we just keep looking until there's no place else to look," said the soft-spoken New Englander.

But where? I asked myself. The sun was rapidly dipping below the horizon. I knew that if I saw one more sunset without finding a likely shipwreck, then the expedition would be over. The Discovery Channel had given us just two full days to find a shipwreck, no matter how hard I had argued for more. It was almost an impossible task.

I didn't want to come all this way and fail. I bit down on the mouthpiece and turned to swim one more row across the harbor. I was swimming from shore now, not a dive boat. I wanted to make sure that I covered the entire bay in my search for the lost ship, and that included even the shallow areas that were close to shore.

Sliding into the water from the cement dock, I dropped to the shallow bottom of the harbor and followed a compass line to make sure I stayed on course. The water was cloudy and the bottom was covered with long-spined sea urchins, making the harbor floor a natural pin cushion. Suddenly I found myself facing a mountainous pile of stone. *I don't remember seeing this before,* I thought. As I rose over it, I could see that the rocks in the pile were all roughly the size of small melons. Their uniform size and shape made them seem out of place, like a pile of rocks in the middle of Madison Avenue. My heart quickened.

I swam over the mound and could see that it was longer than fifty feet and about fifteen feet wide. Covering it were broken pieces of china, mostly blue-and-white porcelain cups and plates. Could this be porcelain stolen by the pirates? I remembered the 1722 account of a voyage to Sainte-Marie by a British naval squadron. They had sent a boat ashore to see what was left of the pirate community and wrote of finding piles of china strewn about on the shore.

I rose to the surface and raised my face mask. On the shore David

Conover, the Discovery Channel producer, had set up a camera and was filming the harbor. He seemed surprised to see me suddenly appear.

"I think we've got a pirate ship here!" I shouted to Bob Paine, who was standing just behind the cameramen. "Bring a metal detector out here. I think we've got it!"

I wanted to put my hands up and holler for joy, but I restrained myself. I may have discovered a ridge of naturally occurring rock protruding through the harbor floor, or a sunken gravel barge. As happy as I was, something told me to wait for further evidence before declaring more than I already had.

Paine swam to the mound and then began sweeping the round stones with the detector's coil. Through the communication phones attached to our masks, I could hear the device indicating that the area he was sweeping was full of metal. Paine circled the base of the mound and saw the unique shape of an oarlock. These would have been plentiful on the *Adventure Galley*.

Paine marked the spot where the oarlock lay in the ballast mound so he could return it there later after it was catalogued. Then he brought it to the surface so de Bry could begin the dating process. We planned to return all the artifacts we removed for inspection from the wreck site. The site would be kept as intact as possible for the Malagasy government.

When Paine and I arrived with the oarlock, de Bry's eyes lit up. "You found this in the ballast mound?" he asked.

"On top of it," answered Paine.

Paine produced a number of large pieces of porcelain from a carry bag, including one cup that was nearly whole. He had marked the location of these artifacts with wire flags so they could be returned later to the exact spots where they had been found.

"Amazing," said de Bry, lost in the object's floral design. "And it's similar to the porcelain we're finding here on land. This is a good sign."

The report of the 1722 British squadron confirmed the presence of

EXAMINING PARTS OF AN OARLOCK FOUND ON THE WRECK SITE. *Paul Perry*

china on the pirate ships, and de Bry reasoned that we were likely to find large amounts of it underwater at the ships' final resting place.

"Was there a lot of porcelain on the ballast mound?" de Bry asked.

"Like pepper on eggs," said Paine.

De Bry thought a moment as he looked at the oarlock and the broken pottery. "Congratulations," he said, holding out his hand. "It looks like we might have found our ship."

Word spread quickly, and soon we were all patting one another on the back, commenting on our good fortune. Everyone gathered around the artifacts, staring at them like they were objects from another planet.

Spiegel pushed his way to the front and slapped me on the shoulder. "This is déjà vu all over again, isn't it?" he declared, referring to the years he had spent working on the *Whydah*.

It was, but it almost seemed too easy. Just over fifteen years ago I had found the *Whydah* in Cape Cod, after looking for that famed pirate ship for years. Now, eleven thousand miles from that site, I was in another famous pirate haunt about to excavate another pirate ship. By comparison, this one was feeling like a piece of cake. Not only had we found it in a matter of hours, the site was close to shore and in a well-protected harbor. We would have no cold winters to keep us out of the water, no storm surges or rough surf to battle as we searched for artifacts.

"Listen, listen," I said, silencing the excited crew. "I think this is it. We're about to go to work on another pirate ship."

THAT EVENING DE BRY began measuring and dating the porcelain. Without specialized reference books it was impossible to establish precisely the age of the fragments. However, certain attributes of the porcelain supported the idea that it was from the Kangxi period (1662–1722). The shape of the foot of one cup, for instance, was a dating clue. So too was the floral pattern—simple two-dimensional drawings—that covered the pieces' pearly white exteriors.

"All of this tells me we're in the right period," said de Bry. He was less certain about the oarlock. He had never seen one from an English galley before and didn't know whether this was the right size or shape, or even the right material. This object was so unique that we were unable to find anything comparable—or any data about seventeenth-century oarlocks for that matter. After some checking, de Bry later found that the oarlock had not been invented until the early nineteenth century. In the time of Kidd, oarlocks consisted of a grommet and pins.

He measured and photographed it and then set it aside to be returned to the wreck site the next day. As soon as he could—which wouldn't be until this first expedition was over—de Bry would take his photos and data to European experts for their opinions on what we had found. In the meantime, he would continue to gather data, interpret what he could, and prepare for the next day's dive, which was certain to present a cornucopia of artifacts.

By ten o'clock that night I was ready to report to the producers at the Discovery Channel. Paul Perry was carrying a satellite telephone from Discovery on which to file Internet dispatches for the company's website, forgoing the primitive and unreliable Internet in Madagascar.

He opened the lid of the case carrying the telephone and aimed it straight up and slightly north. The lid functioned as an antenna that picked up signals from military-communications satellites positioned almost two hundred miles above the earth. The speaker on the telephone began to growl as he adjusted the lid and the connection became stronger. When the squeal reached its peak, Perry pushed a button that locked the telephone onto the signal. I could then make my call.

"Mike," I said to Mike Quattrone.

There was a long wait while the signal went around the world and across seven time zones. Then a voice came back. "Barry. How's it going?"

"It couldn't be better," I said. "We have a ship, and I think it's the *Adventure Galley.*"

I SOUNDED MORE OPTIMISTIC than I should have. We weren't 100 percent sure that this first find was truly the *Adventure Galley,* or even another pirate ship. All we really knew was that we had found a ship in the spot where we expected the *Adventure* to be. And because of the porcelain's date, we knew that the ship we had found was approximately three hundred years old. But we didn't know how many other ships with the same profile had sunk in this harbor. There may have been wrecks that weren't recorded, just as there might have been wrecks that didn't show up in the satellite photos.

Naturally it was the topic of dinner that night. We decided to eat in town at an open-air restaurant called La Bigorne. Maximo Felice, the Italian dive-shop owner, had recommended the rugged little bistro for its fine food and pleasant ambiance.

The mood at the table was festive. Almost everyone ordered bottles of Three Horse Beer, the potent local brew, and began to relax about the strict deadline that had been imposed on us. I was elated, too, but still anxious. "Do you really think we have the *Adventure*?" I asked de Bry.

"I don't know, Barry," he said, sounding cautiously optimistic. "I won't know for sure until I'm down there and take a closer look."

"What could it be if it's not the *Adventure*?" asked someone else at the table.

"Let's put it this way," said de Bry. "It's in the right spot to be the *Adventure*. It seems to have the right cargo. But that doesn't mean it is the *Adventure*. Archaeology is filled with great finds that are different from what they were supposed to be."

We all knew that was true but we were optimistic. After all, in a harbor this size, how many targets with this profile could there be?

EARLY THE NEXT MORNING all of us were in the water, eager to take a close look at the supposed *Adventure Galley*. The water was surprisingly clear, and currents the night before had swept the wreck clean of sediment, revealing the magnitude of what was down there. Blue-and-white porcelain literally covered the round river rocks of the ballast mound. As we peered into crevices, we could see so many shards inside the mound that it looked like someone had tipped over a china cabinet. I swept my hand to remove some of the sediment and revealed an empty but almost intact rum bottle stuck between two rocks. Sail fasteners and other ship fittings were embedded in the mound along with a large cannon that de Bry estimated fired a twelve-pound cannonball.

As we moved around the bottom of the mound, we could see wooden ribs from the ship's hull. They were encased in mud, which protected them from sea worms and other organisms that would have ordinarily eaten them.

We began sticking wire flags in the ballast mound next to artifacts, which we brought up for further inspection. Once they were photographed and measured, the artifacts were returned to their exact spot. Our plan was to keep the site entirely intact until Charlie

BARRY CLIFFORD SHOWS A PIECE OF BROKEN PORCELAIN, ONE OF THE
THOUSANDS OF SHARDS THAT PEPPERED THE WRECK SITES.
Nick Caloyianis

Burnham could create a mosaic photograph of the harbor bottom.
With such a document, we would have a visual record of the way the
wreck site looked when we found it.

One by one the artifacts were brought to the surface and placed in
containers in Felice's dive boat. Later that evening back at the bungalows, de Bry would photograph and measure the objects before returning them to the wreck site the next day.

By now I was 80 percent sure that we were on the site of the
Adventure Galley. One piece of data would help make the identification
certain. When the *Adventure* came into port that day in 1698, it was towing a smaller Dutch ship, the *November*. She had originally been called
the *Rupparell*, but Kidd's crew decided to rename her for the month of
the year that she was taken. Her modest cargo of drugs, clothing, sugar,

59

and coral was sold in the harbor city of Caliquilon in southern India, where her crew was dropped off as well. There she was put under the command of one of Kidd's trusted seamen and brought to Île Sainte-Marie with the *Quedagh Merchant* and the badly limping *Adventure.* Once it got here, the *November* was looted and burned, probably at the same time as the *Adventure.*

To find a smaller ballast mound nearby would almost certainly prove that this site was the *Adventure Galley.*

While the crew continued to work on the first ballast mound, I began searching for a second. Part of the area that I planned to look in had already been searched when we "mowed the lawn." I decided to focus on the area to the northeast, a small patch of harbor where we had not yet searched.

I took a compass bearing from the middle of the ballast mound and began crisscrossing an area that was about thirty yards wide and one hundred yards long. The tide was out and the water was clear. I swam smoothly, just inches over a carpet of spiny sea urchins and tubular worms that writhed on the ocean's bottom. After about an hour of swimming back and forth, I found another ballast mound only fifty-five feet from the center of the first one.

I swam to the surface and saw Paine changing tanks in the dive boat. "Hey, Bob!" I shouted. "I have another one here!"

Testing his regulator, he flipped into the water and swam to my side. In just a few seconds we were down at the ballast mound. Swimming around it, we estimated the length to be about twenty-five feet, less than half as long as the first mound. Protruding from one side were the ship's wooden ribs, covered with coral. There was little in the way of cultural material on this mound, indicating that the ship had been thoroughly stripped before being sunk.

Between the two ballast mounds—the suspected *Adventure Galley* and the suspected *November*—we found a cannon lying on the harbor floor, almost as if it had been dropped there during the transfer between the two ships.

I surfaced and called de Bry over to the site. Everyone else followed as we measured and examined the second ballast mound, and we spent the rest of the day compiling evidence of our find.

That night, over an appalling dinner of green chicken and dirty rice, de Bry laid out his argument for this being the final resting place of Kidd's ship.

"There are things I am one hundred percent sure of," he said. "I'm sure the area next to the dock is the careening area. It's sloping and sandy like none other; then there is no other place that could be the careening area. And we know that Îlot Madame is the area the pirates used to work on their ships. It is referred to by the witnesses at Kidd's trial and also by the British expeditionary force that came here in 1722. It is also marked on a 1733 map."

"Keep going," I said.

"Then there's the context of what we've found," he said. "We've found the ballast mound of a ship that sank accidentally and in a catastrophic way. This was not a ship that was emptied of everything and pulled up on shore. This was a ship that was left out in the water and burned. We know that because some of the ribs we uncovered show signs of having been burned."

"They revolted against Kidd, so they burned the ship right there in the careening area," I said. "That's what the court records told us."

"Right," said de Bry. "Then there's the oarlock. And the ballast mound is covered with broken artifacts, which looks like they left in a hurry and broke a few things along the way."

I could imagine that final day on the *Adventure,* when the angry crewmen stormed the ship to strip it of anything valuable.

"And there's a final piece of evidence," said de Bry. "This might be the most important piece of all: the smaller ballast mound that Barry found this afternoon."

I felt a thrill in the pit of my stomach, like I was on a roller coaster climbing that first steep stretch of track.

8

MURDER AND PIRACY

KIDD SAILED FROM LONDON TO NEW
York to bid good-bye to his wife and chil-
dren. There he waited as long as he could
before leaving for the Indian Ocean, cast-
ing off on September 6, just ahead of the
Atlantic's winter storms.

Kidd, an excellent seaman, plotted a
course that would get them to the Indian Ocean quickly. Thanks to his
careful planning and strong winds that pushed them, they made it
southeast across the Atlantic's dangerous middle stretches without con-
fronting high seas or the doldrums, those windless areas in the sea that
can leave a ship drifting for days or even weeks.

The first stop was Madeira, known to many sailors as the "Island of
Sweet Wine." It was here that the *Adventure Galley* docked on October 8
to take on fruit and alcohol. Eleven days later they dropped anchor at
the Cape Verde Islands to take on salt and water. By the end of the
month, Kidd ordered that the ship set sail for the tip of Africa, the
Cape of Good Hope, beyond which lay the Indian Ocean.

Just before reaching the Cape, Kidd encountered four English men-
of-war under the command of Captain Thomas Warren. Since leaving
England in May, Warren and his four ships, the *Windsor,* the *Tiger,* the
Vulture, and the *Advice,* had been accompanying a group of merchant
ships from the East India Company. They were also hunting pirates.

The merchants were supposed to stay under the navy ships' protection for as long as possible, but the company captains tired of the navy's leisurely travel methods. Nightly dinners became a bore for the merchantmen, who considered them a waste of precious time and a potential danger as well, since one was never sure when a storm might kick up on the high seas. With the death of Warren's sailing master and an increase in scurvy among their own men, the company captains began to grumble.

Scurvy is a nutritional disease caused by a lack of ascorbic acid, or vitamin C, found in fruits and vegetables. After about six weeks without fresh food, the sailors became listless and lost their appetites. Then blotches and sores developed on their skin, their gums rotted, and they lost teeth. This was followed by fever, weakness, and death.

On July 5 the first sailor died of the dreaded disease and the East India Company captains decided to leave the safety of the navy ships for the shores of Africa, where they could lay in a supply of fresh foods and then be on their way.

Thus abandoned by the merchantmen, Warren sailed his fleet to the coast of South America, eventually ending up in Rio de Janeiro. By the time he arrived in this Brazilian mecca, sixty-eight sailors had died from scurvy and many more were dangerously close. Warren let his men recuperate from August until November, when he set out again for the Indian Ocean.

Though Warren's mission was to protect the ships of the East India Company from marauding pirates, the fact of the matter was that he had barely enough men to run the ships in his squadron much less to pursue marauders. He had left England with far fewer sailors than he needed. Now with the death of sixty-eight men, he was desperately searching for new crew members, and it was around that time that he came upon the handsome profile of Kidd's *Adventure Galley*. As an English naval captain, Warren could force a merchant captain to surrender half his crew, and impress them into the service of the king.

But Captain Warren's squadron was anything but a welcome sight to

Kidd. He had lost crew once before to the navy, and if half of his crew were to be impressed out here on the high seas, his mission would likely fall apart.

As the four ships came closer to the *Adventure Galley*, Warren hailed Kidd and ordered him to come aboard his flagship, the *Windsor*. Kidd brought with him a copy of the commission he had received from the king, hoping that its great seal looked impressive even to a manpower-hungry captain like Warren.

The naval captain didn't move on Kidd's crew, at least not right away. Rather than taking hasty action against a representative of the king, Warren decided to feel out the situation over a few days and see where his intuition would take him.

In his short conversation with Captain Warren, Kidd found out that the ships of the line were in desperate need of healthy sailors. Warren made it clear that he would like to have as many as thirty sailors from the *Adventure,* a fair number to give to the king, he argued, without crippling Kidd's expedition. Kidd did not agree, but he continued to sail with the warships simply because he had no other choice.

"You will sail with us to the Cape," said Warren in a voice that made it sound as much like an order as a request.

Against his true feelings, Kidd agreed, returning to the *Adventure Galley* and falling in line with the flotilla.

As he had with the East India Company ships, Warren would sail all day and then gather the ships at night so the captains could join together for dinner. More days spent at sea would buy Kidd some time, he hoped. Once the fleet reached the Cape Colony, it was likely that Warren would force many of Kidd's sailors into the service of the king. So it was best, thought Kidd, that they stay at sea as long as possible in hopes that weather conditions would conspire to give him some kind of advantage for making an escape.

Kidd ate daily with the officers' mess aboard the *Windsor*, displaying his "Vain glory," as Warren described it. During one of these meals, Kidd asked the captain to loan him a new mainsail, since his was in need of repair. When Warren said they didn't have one to spare, Kidd declared that he would have to take one from the next ship he spotted.

Further meals began to reveal what may have been Kidd's true intentions in the Indian Ocean all along. He arrogantly told Warren and the other officers that he would take whatever ship he wanted, not just those of known pirates and the French. In the midst of one of these meals, "disguised with drink," as one of his men put it, Kidd promised a surprised Warren that he could have "twenty or thirty" of his men.

This was most certainly a ploy on Kidd's part to lessen Warren's vigilance. He was still planning to escape, and for that he needed the naval captain to be unsuspecting.

It was right after such a dining and drinking session that Kidd's luck turned. Leaving the officers' mess on the *Windsor,* Kidd noticed that the wind had come to a halt, with all ships dead in the water. Returning to the *Adventure,* he quietly roused his crew, ordered them to break out the oars, and row the ship away from the English.

Kidd and his crew escaped.

BY THE TIME THE *Adventure Galley* reached Tulear on the southern coast of Madagascar on January 27, 1697, word had reached the port that Kidd was a pirate. This was the first official report calling Kidd a pirate, but clearly it was more a rumor than anything else. Warren, obviously stung by being outfoxed, was turning Kidd's dinnertime conversation into a statement of intent. His information wouldn't have stood up in court, since Kidd and company had not yet robbed a single ship.

Kidd heard about Warren's accusation through the captain of a sloop called the *Loyal Russell* from Barbados. He had sailed to Madagascar to buy slaves and fallen ill, possibly from malaria. When Kidd generously transported him to his own cabin for medical attention, he revealed the piracy accusations against Kidd.

After the captain of the *Loyal Russell* had died, the ship's owner asked Kidd if he could accompany him until he got his bearings. Kidd agreed, and they prepared for their trip together.

By the end of February the crew of the *Adventure* had recovered from its long voyage to Madagascar. Fresh fruit and vegetables had cured their scurvy, and the fresh water of the Red Island, a name for Madagascar, had revived their spirits. It was time to head up the east coast of Africa to the waters of rich merchant ships and hungry pirates.

Before his departure for the Red Sea and the Indian coast, Kidd decided to stop at the island of Mohilla, below the Comoro Islands, to careen his ship. The ship was hauled out of the water and tilted on its side with the help of the *Loyal Russell* crew. Kidd's sailors then scraped the marine life off the bottom and filled the seams with caulking to make the hull watertight, a job that generally took at least two weeks.

Unexpectedly, however, many of the men of the *Adventure Galley* began to fall sick. What had started as an ordinary task on a beautiful island turned into a nightmare as Kidd's men died, one after the other. The exact cause of their death is unknown. Some have speculated that they contracted the disease that killed the captain of the *Loyal Russell.* Ultimately thirty of Kidd's crewmen were buried on the tiny island in the course of the next five weeks. By the time the *Adventure* was careened and back in the water, Kidd was critically short of sailors himself.

Kidd left the *Loyal Russell* to its own devices and returned to the island of Johanna (now Anjouan) in the Comoros, where he began recruiting. Several French and English sailors not only joined the crew but also loaned him money to buy supplies.

Historic references suggest that these new sailors were pirates who had made a number of successful cruises and were ready to try again. Clearly the addition of veteran pirates changed the mood on Kidd's ship and the intentions of her crew. Where the old crew might have been willing to adhere to the original agreement with the king and take ships only from countries unfriendly to the Crown, the new crew was in it for gold, no matter what its origin. And everyone wanted reward for their efforts, the sooner the better.

Up at Babs Key in the mouth of the Red Sea they fell in with a fleet of merchant ships and ran up a red flag to signal their aggressive intent. There was only one East India Company ship in the convoy; the rest were Moorish and Dutch vessels. Since the Moorish ships may have been carrying French passes, they would have been legitimate targets for a privateer like Kidd.

Kidd singled out a merchantman, the *Sceptre,* that was lagging behind the convoy. She was a large ship of the East India Company, heavily laden and sluggish in the light wind. It is not known for certain whether or not Kidd knew she was a company ship.

With a lighter, more maneuverable ship, Kidd had the advantage. Or so he thought.

Seeing Kidd approach, the ship's captain, a Scot named Edward Barlow, launched his boats and had his ship towed forward. Suddenly they changed directions, rowing their heavy ship *toward* the *Adventure.* Before he got into range of Kidd's ship, Barlow ordered his cannons fired and his men into the rigging of the ship, where they yelled threats at Kidd's sailors.

Frightened by the sudden aggressiveness of his prey, Kidd moved away to stay out of range of the merchant ship's thirty-six cannons. This game of cat and mouse continued until early evening, when Kidd unfurled his canvas and sailed away.

It was a frustrating encounter for Kidd. He had likely hoped for a Moorish ship, known to carry large sums of money on board. When it became clear that he was attacking an English merchant ship, Kidd was certainly aware that he was about to cross the line into piracy.

AGAINST THE WISHES OF his crew, Kidd ordered the *Adventure* to sail for the coast of India. The ship was leaking badly now, and the water supplies were dwindling. But Kidd insisted that the *Adventure* would have better luck scanning the coast of India for ships to rob than they would on the shipping lanes of the Indian Ocean. It was also easier to find fresh water on the Indian coast.

The crew was angry with Kidd, and they took out their rage at the first ship they stopped, a small English trader sailing from Bombay. They fired a shot across the trader's bow and then pulled alongside as she hove to. The captain, Thomas Parker, went aboard the *Adventure* to try to talk his way out of his problems. While he talked to Kidd, several of the dissatisfied sailors from the *Adventure* climbed aboard the captured ship and began to systematically abuse Parker's men. The lucky ones were beaten with the flat sides of cutlasses. The unlucky ones had their hands tied behind their back and were hoisted aloft so their shoulders were pulled painfully out of joint. All the while, Kidd's crew questioned their captives unmercifully to see if there was any gold hidden on board.

For their cruelty, Kidd's pirates found one hundred pieces of eight. Kidd kept Captain Parker on board as a pilot, and one of the Portuguese sailors was kept on as a "linguistor." The rest of the crew was let go with their ship.

The *Adventure* raised her sail and pressed on to India.

It was here that the official dossier against Kidd started. On September 3, with much bravado, he sailed the *Adventure* into the port of Karwar south of Bombay and filled his water casks under the noses of the East India Company agents assigned to this distant port. Having already heard that Kidd had attempted acts of piracy, the company agents began to question the crew when they came ashore for fresh supplies. Two of the agents, William Mason and Charles Perrin, went out to the ship to talk to Kidd.

Mason knew Kidd quite well. He had led the mutiny against him that took place on the *Blessed William* in the Caribbean, and since then Mason had commanded a ship named the *Jacob*. On a cruise to India a rift of unknown origin had divided the crew and things eventually became so serious that Mason jumped his own ship in Karwar to become an agent for the East India Company.

A veteran sailor in these waters, Mason knew many of Kidd's new crew members. None seemed happy to be on the ship. While Mason waited for Kidd, several crewmen approached the former captain and suggested that they would help him take the ship from Kidd, if he liked. They weren't happy with the way things had gone so far, and they certainly had not taken to Kidd himself.

Kidd appeared on deck, obviously surprised to see a former crewman and enemy from his days in the West Indies. He began telling Mason about his unsuccessful search for pirates, neglecting to mention the incident with Parker's ship and the bizarre exchange with the *Sceptre* at the mouth of the Red Sea. Indeed, Captain Parker was imprisoned below deck as the two men spoke.

Mason noted the shabby condition of the vessel. He found out later from several men who managed to jump ship that the hull was leaking badly and the supplies were down to less than a month's worth of food. He also found Kidd's men to be disrespectful of their captain. In secret they informed Mason that Kidd was both cruel and dictatorial, an unfair commander who played favorites and was harsh in his punishment. Morale was low, and many of the men had taken to fighting with one another out of their frustrations with Captain Kidd. Both ship and crew were falling apart.

As Mason wrote later in the company file that would eventually grow into part of Kidd's indictment:

> Kidd carries a very different command from what other pirates use
> to do, his commission having heretofore procured respect and awe,
> and this being added to by his own strength, being a very lusty man,
> fighting with his men on any little occasion, often calling for his pis-
> tols and threatening any one that durst speak of anything contrary
> to his mind to knock out their brains, causing them to dread him,
> and are very desirous to put off their yoak.

One sailor who did "put off [his] yoak" in Karwar was Benjamin Franks, a New York jeweler who now asked Kidd if he could go ashore on leave. Kidd refused, not trusting him to return. Franks persisted, offering Kidd a beaver hat for the pleasure of shore leave. The bribe worked, and Franks was rowed ashore, never to return again. He later testified against Kidd in a deposition for the East India Company in Bombay, telling them about the attempt to rob company ships in the Red Sea and about the attack on Parker's vessel.

The deposition became part of the file that would eventually lead to Kidd's prosecution.

KIDD AND HIS CREW—what was left of it—had now been at sea for one year. In that time they had taken one ship and netted a total of one hundred pieces of eight. Everyone on board was becoming edgy with the situation and desperate for booty.

In Karwar, Kidd had learned that a large Moorish ship was headed to India from the south. He set out to intercept it. As the *Adventure* left port, word spread down the coast that Kidd was on the prowl.

Two Portuguese men-of-war were sent out to meet him. Both gave chase but only one came close to Kidd's ship. Musket shot and cannon fire injured ten of Kidd's men, but the *Adventure* managed to escape.

The search for the "Moor" ship continued. Sails were sighted and the *Adventure Galley* bore down on them, only to discover that they belonged to the *Thankfull,* an English merchant ship. Kidd questioned the captain but to the disdain of his crew did not take the ship. Kidd appeared to be experiencing a moral dilemma about taking English ships. Apparently he could justify the robbery of small vessels, like Parker's, but not of larger vessels, whose seizure would prompt more attention from the authorities. At least some historians believe that Kidd deluded himself into believing that he was not a pirate at all.

The ship's hull was in need of repair now, and supplies were once again very low. Kidd decided to make for the Maldive Islands to careen the ship and search for more food and water.

Fed up with their captain, the crew unleashed its frustrations on the poor Maldive natives. According to the later testimony of several of Kidd's men, Kidd stormed the island with most of his crew. To establish his authority, he ordered a native tied to a tree, then had one of his men shoot the man dead while the crew ransacked the tiny houses and burned them to the ground. Boats were plundered, and the men turned on the native women, raping them at will. Villagers were powerless to do anything but watch. Other testimony indicated that the atrocities were prompted by the murder of the ship's cooper.

The local men had been forced to make repairs to the *Adventure*'s leaky hull. They made these repairs using myrrh, an aromatic gum. It was a very expensive substance but the only one they could find that was resinous enough to fill in the leaks between the hull planking. Shortly before leaving the island, the ship's cooper made the mistake of going to shore alone. He was surrounded by natives and attacked. Kidd's men found him later, his throat slit.

THE *ADVENTURE GALLEY* MADE its way back out to sea. Hungry for booty, Kidd sailed for the southern tip of India. There they encountered the *Loyal Captain,* an East India Company ship heavy with cargo. Kidd hailed the ship and then boarded her, where he found himself facing Captain Howe, a very nervous veteran of the merchant service who had heard that Kidd was lurking in these waters.

Kidd questioned Howe closely while his men waited on the *Adventure,* ready to stage an attack. In the end, Kidd returned to his ship and let the *Loyal Captain* continue on its way.

Again Kidd's men were shocked and angered, clearly feeling that they were pirates and no longer needed to observe the formalities associated with being privateers. Still Captain Kidd held fast to his change of heart about attacking ships friendly to the Crown. As Captain Howe quickly raised sail and pulled away, a tremendous argument ensued on board the *Adventure.*

The crew took a vote about whether or not they should mutiny and attack the fleeing ship. Most of the crew, "the bigger part," said crewman Abel Owens, was for taking the ship. This raised Kidd's ire.

"You that will take her, you are the strongest, you may do as you please," declared Kidd. "If you will take her, you may take her; but if you go from aboard, you shall never come aboard again."

He thus kept the crew under control for the time being, but their resentment smoldered. They had "gone on the account," a euphemism for turning pirate. By taking Captain Parker's ship and keeping the captain himself imprisoned below deck, they felt they had already established themselves as buccaneers. There was no halfway in this game as far as they were concerned.

For the next couple of weeks Kidd's crew gathered in small groups, voicing their dissatisfaction with their captain as they scanned the horizon for the sails of possible prizes. When Kidd appeared on deck, these small groups evaporated, their participants looking furtively at the now hated captain.

Kidd felt the icy disdain of his men. He felt it in the body language of the sailors, he saw it in the small huddles of crewmen that dispersed in his presence, and he heard it in the conspiratorial words that penetrated the bulkheads and burned in his ears. Kidd was becoming increasingly paranoid, probably with good reason. His crew was ready to get rid of him.

Had the *Adventure Galley* been a true pirate ship, Kidd would probably have been voted out already, since pirates democratically elected their captains to protect themselves against the type of tyranny of the merchant and naval captains that they had left behind. The only time a pirate captain had full control of the men on his ship was during battle. Otherwise, he could be replaced by a simple majority. As the days dragged on, the voices of dissent became louder and louder.

ON OCTOBER 30 KIDD was below deck when he heard the familiar voice of William Moore, the ship's gunner. As he had been for weeks now, Moore was criticizing the captain's decision to free the *Loyal Captain*. Kidd listened to the ridicule, and as he did his anger suddenly soared. Coming onto the main deck, he saw Moore grinding a chisel, surrounded by a group of sailors. Without missing a beat, Moore

IN A FIT OF ANGER, CAPTAIN KIDD MURDERS WILLIAM MOORE.
Artist unknown

directed his argument at Kidd, who, as one crewman described it, "was in a passion."

"Captain, I could have put you in a way to have taken this ship and been never the worse for it," said Moore.

"Would you have me take this ship?" questioned Kidd, his anger peaking. "I cannot answer it, they are our friends. How will you do that?"

"We will get the captain and men aboard," said Moore.

"And what then?" demanded Kidd.

"We will go aboard the ship, and plunder her, and we will have it under their hands that we did not take her," said Moore, whose real plan involved robbing the *Loyal Captain* and then sinking her to avoid being caught for the crime.

Not wanting to take sides in the argument that was brewing, several of the sailors pretended to have duties elsewhere. They remained within earshot, though, and were able to report the following events later at Kidd's trial in England.

"This is Judas-like," said Kidd. "I dare not do such a thing."

"We may do it, we are beggars already," said Moore.

"Why?" asked Kidd. "May we take this ship because we are poor?"

The argument continued, and Kidd's words now became withering. "You are a lousy dog!" he shouted.

"If I am a lousy dog, you have made me so," declared Moore, his anger rising. "You have brought me to ruin and many more."

With that declaration Kidd picked up a heavy wooden bucket, "bound with iron hoops," and swung it savagely, striking Moore on the right side of the head.

The wound was small but lethal, declared Joseph Palmer, a seaman from New York who bent down to examine the gunner. When he inspected the wound, Palmer reported, he could feel "the skull give way."

As Moore was taken down into the gun room, Kidd remained on the deck and proclaimed loudly that the gunner was "a villain." "Damn him!" said Kidd, pacing the deck no more than eight feet above the dying Moore.

Ship's surgeon Robert Bradenham was called for immediately, but there was little the chronically drunk doctor could do.

"What happened?" Bradenham asked as he bent over the stricken man.

"I am a dead man," said the reeling Moore. "Captain Kidd has given me my last blow."

WITH THE DEATH OF Moore the following morning, Kidd now had added a charge of murder to those of piracy that the East India Company was compiling against him. Of those, the issue of piracy apparently concerned him the most. "I do not care so much for the death of my gunner as for the other passages of my voyage," Kidd told Bradenham. "For I have good friends in England that will bring me off for that."

9

THE FIRST BANK
OF KIDD

WE WERE EAGER TO BEGIN A CLOSER examination of the two ballast mounds. But before doing that, we had to create a picture of the entire site, one that would serve to map and record the shipwreck the way we had found it.

Doing this involved swimming over the mounds with video cameras, taking footage the way an aerial surveyor would by using a plane. The map would be created by electronically knitting together thousands of video images into a mosaic picture, a task that would fall to Charlie Burnham.

Imaging specialist Burnham and I have worked together since the early days of the *Whydah* find off Cape Cod. Quirky and humorous, Burnham was schooled in computer sciences at Yale. His regular job was producing videos for a variety of major corporations, but in his "other life," Burnham probed shipwrecks with ultramodern equipment. He has made mosaic maps of some of the world's most difficult underwater sites, including the *Titanic*, which required tens of millions of images shot from the French research submarine *Nautile*. Using those pictures, he helped piece together a complete picture of the world's most famous sunken luxury liner. At one point the *Nautile* acci-

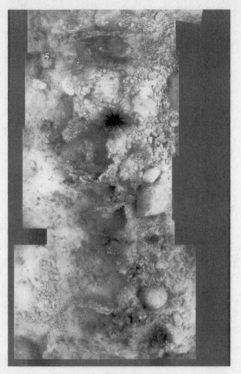

MADE FROM MORE THAN ONE MILLION DIGITAL IMAGES, CHARLIE
BURNHAM'S MOSAIC CREATED A THREE-DIMENSIONAL PICTURE OF THE
UNDISTURBED WRECK SITE. UNFORTUNATELY THE WRECK WAS NOT THE
ADVENTURE GALLEY. *Charlie Burnham*

dentally penetrated the liner's engine room, where it came close to snagging on some of the ship's piping and becoming a part of the vessel's tragic lore.

Making a mosaic of our site in the Sainte-Marie harbor carried its own set of problems. The *Titanic* was photographed using an electrical guidance system, an invisible grid that guided him and his camera. For the *Adventure Galley* site, Burnham used a grid of nylon line designed on the spot by Wes Spiegel. Spiegel created it on the sandy beach in front of our bungalows by pushing sticks into the sand and lacing them together with nylon line, creating a small-scale model on land of what he planned to do under the sea.

Then, over the course of several days, Spiegel and Burnham, accompanied by Bob Paine and my son, Brandon, pounded stakes into the ocean floor and joined them with nylon line, which would serve as guides when they scanned the site with their video cameras.

Eventually they would shoot more than one million images of the wreck site. By Burnham's own estimate, putting together the map would take about "one thousand hours," resulting in a map of the wreck site that could be used during the second expedition, when the actual excavation was scheduled to begin. And because the map was digital, it would allow us to see things as small as a tiny sliver of glass, thus letting us plot our excavation in minute detail from the comfort of our own homes back in Cape Cod.

For more than forty years in the late seventeenth and early eighteenth centuries, Île Sainte-Marie was occupied by pirates and those who traded with pirates—an almost unparalleled record in the annals of Anglo-American freebooting. Because of that, plenty of artifacts and sites could be found on the island, some in plain view. Cannons from the pirate era were everywhere, so many that the locals thought nothing of using them for the island's cement dock works, or as fence posts and grave markers.

Broken china could be found at many spots on the island, not just the careening area, a reminder of a time when plates were so plentiful they didn't have to be washed after a meal but were just thrown away or used for target practice. There were sites of forts and other crumbling buildings in which the pirates once lived in tawdry splendor, such as Adam Baldridge's pirate fort on Pirate Island.

Plopped in the middle of the Bay of Pirates like a crumpled piece of paper, this island, as André Mabily had told us, was *fady*, or taboo, to the locals, haunted by the spirits of pirates who once lived there.

Many of the locals offered an emphatic "no" when we asked them if they would go there with us. One nervous elderly man, who had lived on Sainte-Marie since birth, said, "What you will find there you do not want. There are ghosts, and if you see them you will die. It is that simple."

Mabily believed what everyone said about Pirate Island but had nonetheless agreed to lead us to the caves, assembling enough dugout canoes, or pirogues, for about half my men and the Discovery Channel film crew. Several young Malagasies had also been recruited to row us to the island in awkward boats that were nothing more than very heavy hollowed-out tree logs.

After a brief discussion of logistics, we climbed into our respective boats and made the unsteady voyage across the dark waters of Pirate Bay to the lush, forbidden island. We reached shore ten minutes later and immediately found signs of someone who had indeed tried to live there and failed. A short pier jutted from the shore. At its land's end was the foundation of a house and a collapsed grass roof.

"What's this?" I asked Mabily.

"A Korean couple came to the island fifteen years ago and tried to live here for a while," he said. "They planned to raise oysters for pearls to sell them back home. They built holding tanks and buildings to do the work of packing them, but they didn't last very long. They decided to leave."

Other rumors had it that the couple came to the island looking for treasure. Whatever the case, "Sainte-Marie can be hard on outsiders," said Mabily.

I'd noticed the disillusioned looks on the faces of some recent European émigrés who occupied homes in the village, suggesting that this was not the bucolic Eden they'd thought it was going to be. It must have been similar for the Koreans, both on the main island of Sainte-Marie and on their lonely enclave of Pirate Island.

"This island is *fady*, and a person in a *fady* place does not feel welcome. It is the ghosts. You do not have to see a ghost to experience their presence," Mabily said as we began to explore Pirate Island.

MOST PROMINENT AMONG THE local lore about pirates is the story about how they used this strategic island as a refuge against attack. Looking at its steep shores and heavy jungle, one feels the legend is fairly convincing. Pirate Island was a natural fort. John de Bry would later find a map showing an anchor chain across the mouth of the harbor to prevent unwanted ships from approaching the island. A ship trying to attack Pirate Island would be kept out of the bay and become an easy target for cannon fire from both the island itself and the hills that lined its harbor's mouth.

We were intrigued by the obvious reason for the security of this island: this might be where they hid their booty. Many pirates came from mining families in England and Scotland, where they learned tunneling skills as very young men. De Bry and I speculated that the pirates dug tunnels on the island to hide their booty and themselves. If true, the island would have been a place of banking as well as a source of refuge.

We weren't the only ones who felt that way. As we trekked through the heavy brush that led to the first tunnel, Mabily talked once again about Bi Bq, the late grave robber who'd come to Sainte-Marie from Réunion Island in search of pirate treasure. "He came here with an army of hired hunters and dug through all of the tunnels," said Mabily.

"Did he find anything?" someone asked.

"I don't know," Mabily said, continuing a rapid pace up a steep hill. "Bi Bq is dead now, so we cannot know for sure what he found. I do think that he found something and then concealed his path. All of the tunnels are sealed up with cement."

We came to a deep hole in the ground that was obscured by thick underbrush.

"Are you sure?" I asked Mabily.

"I am sure," he said, swinging his machete to clear the foliage. "I followed this one back as far as I could. It was very cold, and then I came to a part that was cemented shut and I could go no farther."

BRANDON *(FAR LEFT)* AND BARRY CLIFFORD FOLLOW THEIR LOCAL GUIDE TO THE TUNNELS OF PIRATE ISLAND. *Paul Perry*

We all gathered around the tunnel and peered into its depth. Brandon grabbed a branch and climbed down into the tunnel. A moment later Eric Scharmer followed him in. They went as far into the tunnel as they could, which was barely out of the reach of darkness, where they encountered a plug of what seemed to be cement. There was no way around it.

Mabily thought that had he gone all the way in, he would have come out at one of the other caves, to which he took us now. We crossed the island and stopped at a thick cropping of brush that surrounded another cave. Once again Brandon jumped into the hole and poked around for an entry point. Finding nothing, he climbed the cave's craggy edge and emerged at the top.

Not far away was another cave, and yet another down by the water's edge. These too were blocked.

JOHN DE BRY *(LEFT)*, BARRY CLIFFORD, AND BEN PERRY FIND BROKEN
POTTERY FROM A PIRATE SHIP ON THE SIDE OF A HILL ON PIRATE ISLAND.
Paul Perry

"I am sure all of these tunnels join inside the island," said Mabily.

We felt the same way. And we wanted to see them. As we drank milk
from a fresh coconut that Mabily plucked from a tree and cracked
open with his machete, we began to speculate as to the nature of the
tunnels.

"I would guess that there are a couple of underwater chambers
down there," said de Bry.

"That would make sense," I said. "Or maybe chambers that could be
flooded. That way they could open a dam and conceal their booty even
more if raided by the British."

"That would be one hell of a vault," said de Bry.

"It's the Captain Kidd International Bank," I said.

10

NOW I AM ONE

KIDD'S COMMISSION FROM THE KING and his agreement with Bellomont and Livingston clearly spelled out that he was to seize the "merchandizes and treasures" from pirates and from countries that were not friendly with England, namely, France.

But the question wasn't as much how to capture French ships—Kidd's large and surly crew could certainly do that—it was how to *find* them. In a sea plied by merchant vessels from England, Portugal, Holland, Spain, and many other nations, the ships of France represented a small percentage of those that the *Adventure* was likely to encounter.

To increase that percentage, Kidd fell back on an old trick. Rather than fly the flag of England, he opted to fly a French flag when approaching a potential prize. It was common in those days for a ship to carry several flags as well as passports and statements of ownership. That way they could fly the flag of the ship pursuing them, making it less likely that they would be stopped. If they were stopped, captains would then produce false papers, providing additional "proof" that they were from a friendly country.

The first ship to fall for this ruse was the *Rupparell*, a coastal trader that was flying no colors. Kidd changed course for the small ship and raised the colors of France, which at the time were plain white. The

Rupparell steered away from Kidd and headed for shore. Kidd moved in closer and fired a cannon shot across her bow and then another. Finally the *Rupparell* dropped her sails and waited for the captain of the *Adventure Galley* to make his boarding call.

To maintain the masquerade, Kidd had a French crewman named Le Roy pose as the captain of the *Adventure Galley*. After a few nervous words were spoken, the *Rupparell*'s Captain Dickers went to his cabin and returned with a French pass. His ship was in fact Dutch, and he likely carried the forged passes of many nations in his cabin. But as soon as he'd presented himself as French, Kidd stepped around his imposter captain and pointed at Dickers.

"By God, have I catched you?" he shouted with joy. "You are a free prize."

As Dickers lamented his fate, Kidd's men inventoried the cargo of their catch. In addition to the ship itself, there were two horses and eleven bales of cloth in the holds. Kidd asked the captured men if any of them wanted to join the crew of the *Adventure*. To Kidd's surprise, Captain Dickers volunteered, as did two of his men. The rest were allowed to row ashore in the *Rupparell*'s longboat.

The captured ship was renamed the *November* by Kidd's crew, after the month in which she was taken.

The *Adventure* centered on Caliquilon, a remote southern Indian port where the meager cargo was sold to Gillam Gandaman, a former employee of the East India Company, in a transaction completely contrary to the terms of Kidd's contract with Bellomont and partners.

The *Adventure* left port and pressed on. Further south they stopped a ketch and stripped it clean. Her cargo of bacon, sugar, candy, tobacco, and myrrh was divided among the crew. Several weeks later a Portuguese ship was plundered and two chests of opium from her holds were sold ashore.

The crew was happier now that they had put their newfound skills to use—and Kidd was agreeable, since most of the victims were foreign—but still they longed for a large take. With the *November* tagging behind, the *Adventure* plied the waters south. Lookouts manned the crow's nests, and sailors on deck scanned the horizon constantly. They were searching for the one ship that would make them rich, a merchantman filled with goods and money and bound for the port of a distant shore.

On January 30, 1698, they found her. As they sailed near the tip of India, white sails suddenly became visible.

"Ship ahoy!" shouted one of the lookouts.

"Ship ahoy!" shouted several other sailors, many of whom hadn't even seen the sails on the horizon. They were hoping to get the extra hundred pieces of silver promised to the first man to see a major prize.

What the lookout saw was the *Quedagh Merchant,* and she certainly was a major prize. Weighing in at four hundred tons, the Armenian-owned ship was loaded with opium, raw silk, calico and muslin cloth, and various other goods including iron and saltpeter. The owners estimated the value at somewhere between 200,000 and 300,000 rupees, equal to £12,000, a fortune in its day.

Kidd steered his leaking ship toward the massive vessel and ran a French flag up the flagpole. The *Adventure* closed on the *Quedagh* and ordered that she heave to for a search. Captain Wright of the *Quedagh* ran up a French flag, too. For a moment the two ships sat motionless like pieces on a chessboard until Kidd made the next move.

"Come aboard!" shouted Le Roy, the decoy captain for Kidd.

Captain Wright sent over his own decoy, a French gunner with the phony French pass of his ship, to pretend that he too was the captain.

After a brief chat, Le Roy made it clear to the nervous gunner that he didn't believe the man was the captain. The gunner returned to the *Quedagh,* and Captain Wright came to the *Adventure* by return rowboat. As soon as he boarded, Kidd introduced himself as captain and claimed the *Quedagh Merchant* as a prize.

The owners of the cargo were frantic. As Kidd waited, the seven Armenians put their heads together and one of them, Coji Babba, offered twenty thousand rupees if the ship were set free.

"That is a small sum," said Kidd. "The cargo is worth a great deal more."

The offer rose to forty thousand rupees, but Kidd would not budge. The cargo alone was worth nearly ten times that much, and the ship herself was becoming more valuable with every new leak that sprung in the hull of the *Adventure Galley*.

Kidd returned to the tiny port of Caliquilon, where he made Gillam Gandaman happy once again by selling him the contents of the newly captured merchant ship. Estimates are that Kidd and his crew grossed anywhere from seven thousand to twelve thousand pounds in the sale. Plus they still had bales of cloth in the holds waiting to be sold at a later date in a better market.

Leaving Caliquilon, Kidd reportedly tried to attack the *Sedgewick*, an East India Company ship. Capturing such a slow and heavily laden cargo vessel should have been easy, but the *Adventure Galley* was now beginning to come loose from the constant motion of the sea and lack of repair. Try as she might, the *Adventure* could not gain on the slow but steady company ship.

Kidd gave up the chase and turned his flotilla to the southwest, with the *Adventure* being pumped day and night by exhausted Indian sailors who had been captured from a passing ketch. They worked the pumps hard but could barely keep up with the water that flooded in through the worm-eaten hull. After a few days of constant pumping in which the galley was getting dangerously low in the water, someone came up with the idea of wrapping the entire ship with several lengths of cable and securing them tightly like a corset. It was an improvement, but the *Adventure* was still struggling to remain afloat and teredo worms were eating the English oak like it was fine cheese. Many landlubbers thought of sharks as the most dangerous denizens of the deep, but to an English sailor in the Indian Ocean, these wood-eating worms were the real nightmare.

Kidd's next stop would be Île Sainte-Marie, more than twenty-five hundred miles away. Could the *Adventure* make it? *If the weather holds,* thought Kidd, *and the pumps keep working, and the cable stays tight, and the Indian sailors don't break their backs. If, if if . . .*

With the tiny *November* being towed behind, and the massive *Quedagh Merchant,* commanded by George Bollen and manned by Kidd's most trusted crew members, following, the flotilla headed for Madagascar. Kidd had ordered Bollen to stay within sight of the *Adventure,* and the loyal sailor agreed that he would try.

In a few months they would be in the world's most infamous pirate stronghold, surrounded by the criminals that his backers had expected him to arrest. Did Kidd see the irony? Or did his ego blind him to his own shortcomings and crimes? Maybe he had a good laugh about all of this out there in the Indian Ocean as he slogged to that distant pirate haven. He had been sent here to capture pirates. Now, although he never admitted it, he had become one himself.

WITHIN A WEEK OF taking the *Quedagh Merchant,* Kidd called his crew together and announced that he was having second thoughts about taking the merchant ship after all. "We should return this ship to Captain Wright and be done with it," he said to his men.

With the capture of the *Quedagh Merchant,* Kidd felt that he had crossed a dangerous line, into a territory from which he could not return. Robbing a small English vessel like Parker's might be overlooked, as would the taking of a Dutch ship, he explained to his crew. But the seizure of the large cargo ship would "make a great noise in England," one that would not be ignored by the law. Kidd knew that the cargo was owned by men with influence at the mogul court who would make trouble—trouble that might be difficult to avoid. He pled his case, making it clear that they had promised the king of England himself that they would not take English ships or those of their allies.

Kidd begged the crew to let him return the *Quedagh Merchant,* but not surprisingly, they roundly rejected the idea. It is fair to say that at this point the crew was more than angry, even threatening. Court documents later revealed that the majority of the sailors were in favor of a mutiny, although none had the courage to carry it out. They most likely backed off because Kidd didn't press his point further. Although he had been adamant about returning the ship, Kidd knew that he was walking a fine line with his crew, running the risk of being killed or spending the rest of the voyage chained below deck.

The tiny flotilla relied upon the summer trade winds of the Indian Ocean to drive it slowly south. They pursued the summer route, straight down from the tip of India, past the Maldives and then west toward the Cape of Good Hope. There is record of Kidd's fleet robbing several ships, mostly small Portuguese traders, on its way to Madagascar. Kidd's booty-hungry sailors stripped these vessels of their cargo, leaving them just enough food and water to reach the Maldives or the Indian coast. The buccaneers of Kidd's flotilla were on a long voyage, they told the sailors of the ravaged ships. They were on their way to the pirate paradise of Île Sainte-Marie.

The men they robbed had heard stories of this pirate island. In their brief meetings all exchanged what information they had. Kidd's men talked about the coast of India, and the captured sailors lauded the island of abundance, Sainte-Marie. Although less than two miles wide and twenty-six miles long, she held everything that a mariner could want. Crops of all kind—rice, bananas, mango, pineapple, and manioc—grew easily on the fertile and rainy island. Pigs and chickens were raised in pens, while zebu and turtles ran wild. Alcohol was brewed from sugar and fruit juice. There was even a special drink called *toke* that was fermented from honey.

The natural beauty of the slender island was matched by that of the Malagasy women, the product of an exotic gene pool that included people of African, Arab, Indonesian, and European descent. Lithe and affectionate, these pleasant island women promised to be a delight for the sea rovers.

Many pirates left the trade after landing on this island. Having made their fortune at sea, they now found it preferable to live like kings rather than returning to Europe. Thanks to the natural protection provided by the coral reef that surrounded much of its eastern side, they were able to do that on Île Sainte-Marie. Impassable to any ship, the reef protected the well-known pirate hangout from any surprise attack by pirate hunters.

The more Kidd's sailors heard about Île Sainte-Marie, the more excited they became.

ALTHOUGH KIDD ALREADY KNEW these stories, the captured sailors also talked about the "white chiefs" of Sainte-Marie, namely, Adam Baldridge and Edward Welch.

Sainte-Marie was originally established as a pirate base by Adam Baldridge, though little is known about this rogue entrepreneur. Some believe that Baldridge murdered a man in Jamaica and became a buccaneer himself in 1685, but that story has never been confirmed. What is known is that Baldridge was on the crew of a slave ship that landed on Sainte-Marie in 1690. Something about the island appealed to him, and he stayed.

To endear himself to the natives, Baldridge helped them repel attacks by mainland warriors. This was a big help, since the islanders were few and frequently preyed upon. Now, with the modern weaponry used by Baldridge and other Europeans who joined him, they became a power to be reckoned with. Not only did Baldridge help repel the invasion of warriors from the mainland, he also led attacks on the mainland himself.

In gratitude for his help, Malagasy chiefs gave Baldridge cattle, land, and two of their daughters as wives. Many would have stopped there, but Baldridge was something of an empire builder. With the willing help of native labor, he built a settlement on the south end of the island. Then he went on to erect a fortress with cannons overlooking the harbor. When that was completed he began marketing his pirate village to merchants and pirates alike.

A John Finlinson testified in 1698 before the Board of Trade that

Baldridge was gifted with strong business abilities and a good sense of diplomacy. Said Finlinson:

> Baldridge had then about twenty whites with him; most of them English. They were a disorderly Crue, always quarrelling with one another. But Baldridge and one [Lawrence] Johnson [his business partner], agreeing together, got the command of the rest. They have ingraciated themselves with the Negroes of the Island who submit to Baldridge as their King. Those Negroes are peaceable people and do not sell any slaves; but Baldridge directed the Master of the Ship in which this Seaman was to a Bay in the Island of Madagascar, about sixty miles distant, where they got what they wanted. Those that live there with Baldridge have plenty of Cattle, and Sell Victuals to Pirates.

With the construction of his pirate village completed, Baldridge wrote a letter to Frederick Philipse, the New York merchant and slave trader. He convinced Philipse that he could more than double the wealthy New Yorker's profits, promising to sell him slaves at thirty shillings apiece. Plus he promised a ready market for any goods that Philipse might want to sell to the pirates on his island.

Philipse liked what he read and sent the fully loaded merchant ship *Charles* to Sainte-Marie. Its cargo was rich and diverse, including such goods as rum and Madeira wine, and "books, Catechisms, primers and Horne books and two Bibles." The goods fetched a whopping sum of eleven hundred pieces of eight as well as thirty-four native islanders and various items.

The next four years saw an expansion of trade on the island, as traders sent their merchant vessels to do business with the pirates. Baldridge built a warehouse to hold the goods sent over by Philipse and the other shippers.

Sainte-Marie quickly became popular among Indian Ocean pirates. It gained the reputation of being a safe and fair place to trade booty, careen a ship, even divide up her cargo. It probably was Baldridge who had a chain strung across the mouth of the harbor as added security for the ships that docked at the island. If his cannons didn't keep an unwanted vessel out of the harbor, the chain would.

Soon the island of Sainte-Marie became so strong that even the English were reluctant to attack it. Captain Thomas Warren, the commander of the flotilla that Kidd had escaped from, wrote in the log of his flagship, the *Windsor,* that the island's population had grown beyond his ability to successfully invade it. "The pirates have a fine harbour where they clean their ships. They have built there a regular fort and mounted 40 to 50 guns. They have 1,500 men with 17 ships and sloops, some mounted with 40 guns. New York, New England and the West Indies send them food and supplies."

Just when Baldridge seemed to have it all, he blew it. In July 1697 he purchased an interest in the *Swift,* a brigantine with a large cargo hold. Baldridge told the island natives that he was going to use the ship as a coastal trader, purchasing goods from the mainland and returning to the island to sell them to pirates.

To celebrate his new purchase, Baldridge threw a party and invited about a dozen Sainte-Marie families on board. They rowed out to the ship that evening in their pirogues. The party swelled to more than one hundred as the pirogues kept arriving, filled with happy islanders.

The "White Chief," as he had come to be known by now, entertained them well. He served large chunks of zebu and put everyone into a drunken stupor with rum, *toke,* and *bumbo,* a potent native treat. Then, as the natives passed out one by one, Baldridge had his men take them down below decks and slip shackles on their wrists. The islanders were going to be sold into slavery.

When news of Baldridge's deceit reached shore, the natives went berserk. They had known that Baldridge was involved in the slave trade because they had seen the ships arrive and had even helped him by participating in slave roundups from the mainland. But they didn't think that the man they considered their white chief could commit such an act of betrayal against his own people.

As the *Swift* sailed to Réunion Island to sell its human cargo, the remaining natives of Sainte-Marie rose against the men Baldridge had left behind. In a savage battle they killed thirty-four of them, including his partner, Johnson.

Baldridge left Sainte-Marie for New York, where he attempted to get Bellomont to lobby on his behalf to reestablish his "settlement." When the Lords of Trade, understandably, rejected that idea, Baldridge left town rather abruptly.

After Baldridge, Edward Welch became the next "White Chief" of Sainte-Marie. Welch knew of Baldridge's plan and had the wherewithal to leave the island before the carnage started. He boarded a ship and waited in the harbor for the islanders to calm down. Then, in an act of great courage, he returned to the island and convinced the chieftains that they still needed someone to conduct the trading business. Forced to make a deal with a devil, the chieftains decided to deal with the devil they knew. Welch had been on the island for six years now, and, at least to some extent, they knew what to expect from him.

All of the facilities had been burned to the ground. The warehouse was little but charcoal when Welch came back from his temporary exile. The same was true of the fine fort Baldridge had built and the tavern where the sea brethren gathered to drink and talk. All was ashes, and Welch had to embark on a major rebuilding project.

He erected a new fortress about four miles north of the harbor, moving six of the big guns from the burned fortress to the new stockade.

In the months after Welch replaced his old boss, the island of Sainte-Marie returned to normal and the men who had survived the uprising settled cautiously into their old way of life. They moved back into their platform houses and nurtured their frightened wives, adopting the wives of their dead companions so that they would not have to live as widows.

Sainte-Marie once again became a port of call that was open for business. Ships arrived to buy and sell. Gradually Welch rebuilt his stock of merchandise. Ship manifests of the day confirm that Sainte-

Marie had a healthy supply of goods, including rum, wine, beer, salt, flints, pistols, knives, looking glasses, combs, buttons, scissors, cotton, thread, hats, shoes, pipes for smoking, needles, European clothing, and Bibles. The residents of Sainte-Marie had everything they needed. They lived happily in a lush, exotic haven.

Then Captain Kidd and his crew arrived.

11

ZEBU-QUE

OUR FRENCH GUIDE, GILLES GAUTIER, had arranged our in-country travel and accommodations, some of which I tried not to hold against him. It was his job to be our "fixer," the person in the know who would keep us from violating any local customs. He had been invaluable in explaining to local officials what it was we were doing, disabusing them of their notions that we were looking for hidden gold or jewels. Now, however, Gautier had encountered something he couldn't fix. The locals wanted us to sponsor a zebu ceremony.

"What is a zebu ceremony?" I asked Gautier when he told me of the bizarre request.

"It is a sacrifice, really," he replied in his rich French accent. "You must buy a zebu so they can sacrifice it to their ancestors. It is a local custom. The earth is very holy to the Malagasy, and you are interested in people who are buried in the earth. Pirates and their other ancestors are buried there, too, so you must show respect by having a zebu ceremony."

"What does a zebu ceremony have to do with people buried in the ground?" I asked.

"You have to kill the zebu so the blood flows down into the ground and appeases the ancestors," said Gautier. "It is to offer the blood of

CELEBRANTS SURROUND THE ZEBU CEREMONY'S CENTER OF ATTRACTION.
Margot Nicol-Hathaway

life to the dead pirates so they will approve of what you are doing. These pirates are their ancestors, and they are what you are looking for, so an act like this will be good for you."

Other members of the crew looked at me and shrugged.

"Since you're in charge, you get to do the honors," said Wes Spiegel.

The honors, I soon learned, amounted to decapitating the zebu with a machete. It all seemed like bloody business, but according to Gautier it couldn't be avoided. "If you don't do this, the people here won't think very much of you," he said.

I stopped to ponder that statement for a moment. I had heard the story of Bi Bq, the outsider the islanders hadn't thought much of and who was mysteriously murdered in his own country. And then there was that red ghost, the one that was fatal to all who glimpsed it. I didn't want him mad at me, either. On this expedition we would need all the help we could get, both natural and supernatural. This would be an important show of respect.

"I guess we have to do it," I told Gautier. "How much will it cost?"

"About three and a half million Malagasy francs," he said. I did the math in my head and emitted a low whistle.

"That's over five hundred dollars," I said.

"Yes, it is," confirmed Gautier. "First you have to buy the zebu. Then you have to pay a butcher to tie it up and sacrifice it. That is, unless you want to do that yourself."

I paid the five hundred dollars, and the next morning we found ourselves watching uncomfortably as a Malagasy butcher and two young assistants struggled to tie up a frightened young zebu in a grassy area next to the wreck site.

It was a cloudy day, and we were standing in a semicircle around the zebu as the tribal elders and dozens of other islanders pressed together for this solemn occasion. But for the fact that the elders were dressed in their finest Western clothing, this seemed to be a ritual from the distant past. One by one they folded banana leaves into the shape of a cup. Then they filled their cups from a Pepsi bottle containing a homemade brew called *betsa betsa*. As they drank, they offered a prayer of thanks to the zebu for giving up its life, hoping that the blood that seeped into the spongy ground would quench the thirst of their pirate ancestors.

Someone folded a cup for me and filled it with the cloudy beverage. I gulped it down in one swallow, and the cup was immediately refilled.

"I wouldn't drink that," said Charlie Burnham. "That stuff is made with local plants that haven't even been identified by the U.S. government. They have botanicals here that could put you into another dimension."

I didn't pay attention to his advice and swallowed the contents of the second cup. Almost instantly I could feel it fill my body and go down to my feet. I was overly relaxed for a moment and then felt as though I was lifting just slightly out of my body. Brandon and I sat by the edge of the lagoon feeling the effects of the mixture and wondering out loud where the *Adventure Galley* lay.

The butcher sharpened his tools and then positioned the animal's head so he could do his deed with the greatest dispatch.

There is no need to describe the struggling and braying that went on as he sliced through the zebu's jugular vein and then the rest of his neck. I will say, however, that this was as close to the realities of the food chain as I have been in several years, and it left me feeling somewhat uncomfortable and queasy . . . or was that the *betsa betsa*? I wasn't entirely sure.

I watched as the butcher stripped the skin from the freshly killed zebu and then began the process of butchering the small animal and handing out large chunks of meat to the natives, who carried it in banana leaves to a barbeque fire that had been started in the playground of the elementary school. There the meat was skewered and placed on a grill.

In one of the classrooms we sat at the students' desks, which had been arranged so our backs were against the walls and we were facing a table in the center of the room. Meat and cooked vegetables, soda and beer began to materialize on the table from islanders whose job it was to stock this gathering.

Before we ate, I stood and presented to the island's mayor a large sum of cash—five hundred dollars—for the Sainte-Marie school system and another large sum for the restoration of the colonial governor's house, a beautiful residence near the wreck site that had been damaged by a fire of unknown origins.

I don't remember what I said in my short speech, but I do remember that the mayor was delighted and surprised that an outsider would make such a generous gift to the island's poor infrastructure. The country's assistant director of patrimony, Limby Maharavo, had come to the island from the capital, and he, too, was impressed by the donation.

"Most outsiders take advantage of this island's resources but don't go out of their way to give anything back," the mayor said in a predin-

As those who know better look on, Barry Clifford downs his second cup of *betsa betsa*.   *Paul Perry*

ner speech. "We are glad to have your expedition in Madagascar and are looking forward to a long-term relationship."

When the applause died down, the assistant director of patrimony stood and expressed his gratitude as well. "Madagascar's history is a mosaic with many pieces missing," he said. "Operations like yours will help fill in those empty spaces."

There were cheers and more applause. At that moment I felt certain that we had overcome any distrust that might harm our relationship with this country's bureaucrats and ruin the possibility of future expeditions. I didn't want them to think that I was going to sell artifacts from their country. In fact, I had been careful to tell both the mayor and the assistant director what it was we were finding on the wreck site. Ours was strictly a mission to survey the sites of the pirate ships that had sunk in the harbor. Later, when artifacts were brought up from the bottom during future excavations, the Malagasy-government authorities would be the first to be advised. I promised all of this to them, and now I had sealed that promise with donations to their schools and to a colonial building that, according to the mayor, would be the site of a pirate museum when it was restored.

"When you're drinking *betsa betsa* with the locals, you've been accepted," I said to one of the expedition members.

It wasn't long before I realized I had been naïve and mistaken—unimaginably so.

AFTER THREE WEEKS ON the island, we returned to the United States to digest our data. In some ways it seemed as though we had too much information to analyze, and in others it seemed as though we didn't have enough.

Burnham had thousands of digital video images that, when knitted together, would provide a clear map of the wreck site. We also had photos and measurements of the various artifacts we'd found, mainly pieces of pottery. De Bry had dated and identified the unique shards as being from the Kangxi period (1662–1722). He was going to send the photos to experts in France who could give him a second opinion.

It looked like we had a lot, but it wasn't everything we needed to make a positive identification on the ships. We kept our eyes open for any glimpse of the ship's bell—the best source of identification, since the name of the ship was often cast on the side. Wood samples would have been a close second. By removing some small splinters of wood from the burned pieces of hull that were trapped underneath the ballast mound, we could employ dendrochronologists in England to provide not only the approximate year that the ships were built but perhaps even the forest where the wood came from. But we were unable to remove wood from the wreck site, since our permit allowed us only to survey the wreck, not excavate it. And although I doubted that the Malagasy government would mind if we took a chip or two of wood, we couldn't risk asking. Such a question might make them think that we'd found gold and were trying to trick them into letting us take it.

"There are not quick changes in Madagascar," said Gautier. "If you ask them to change your permit, they will make you stop the work you are doing while they take several weeks to decide whether you should get the new one. Everything is slow in this country. And suspicious. Be happy with what you have."

And so we left the island with what we had—data and questions— and were happy.

Before we left I hired Gregory, the night watchman at the Orchidées, to keep an eye on the wreck site and told him we would be back in a few months. It was his job to watch the site daily and make sure that no one dove on it or took any artifacts.

Gregory took what we were doing very seriously. He considered our work to be sacred, similar to the Malagasy tradition of *Famanihana,* in which dead relatives are dug up every few years for a party with the living. During this party, which is carried on in conjunction with Halloween, the ancestors impart stories about the family history. "What you are doing with the pirates is similar," said Gregory. "You want to collect stories."

During the four months that we would be gone, I expected Gregory to just look at the site on his way to and from work. I was told later, however, that Gregory was very vigilant in his duties, spending long hours there each day, keeping the site safe for our return.

12

BRETHREN OF THE SEA

AROUND THE FIRST OF APRIL THE ordeal aboard Captain Kidd's ship ended.

"Land ho!" called a crewman in the riggings. Hearty shouts broke the dissatisfaction and monotony of the previous weeks as the sailors strained their eyes to see this dream of coral and tropical vegetation rise over the horizon.

The *Adventure Galley* kept clear of the island's treacherous north end and rolled gently along its west side. Seamen lined the rail on the port side of the ship and drank in the scenery: tropical beaches, coconut palms, and a heavy forest of strange but beautiful trees and plants.

Eventually a small settlement came into view. Then the mouth of the harbor appeared. Then a ship. She was anchored near the careening area and had her guns trained on the harbor's entrance. A dugout canoe had been launched from the ship and was rowing its way toward the *Adventure Galley*. The men in the crew were English, and they had fear on their faces as they boarded the newly arrived ship. They knew from sailors who had come from England nearly two years ago that Captain Kidd had been sent by the king to fight pirates, and they didn't know yet that Captain Warren and other English officials had

THE CORAL-RINGED TIP WOULD HAVE BEEN THE FIRST VIEW KIDD AND HIS CREW HAD OF ÎLE SAINTE-MARIE.   *Paul Perry*

accused Kidd of being a pirate. Nor did they know that Kidd had recently committed several acts of piracy. Amazingly enough, the hub of piracy in the Indian Ocean was behind on the news.

"We have heard that you are here to take us and hang us for being pirates," said one of the buccaneers.

"There is no such thing," Kidd said.

The men in the canoe climbed on board the low and leaking *Adventure,* laughing. They told Kidd that forty of their fellow crewmen had gone to hide in the jungle upon hearing of Kidd's arrival. Now two of the pirates climbed back into their canoe and rowed ashore to tell their compatriots that Kidd was here to join them.

Kidd asked them who their captain was. When he heard the name Robert Culliford, he must have felt a twinge of anger. Culliford had been one of Kidd's crewmen aboard the *Blessed William,* the ship that had been taken from Kidd when he bravely fought the French in the Caribbean nearly ten years earlier. Culliford had joined the mutineers who stole his ship. They were the ones who considered his ego obnoxious and his "ill behavior" intolerable.

If memories of the *Blessed William* were flooding Kidd's brain, he did not make it known to the people he was talking to. He wanted to be taken to Culliford, he said. Boarding the canoe, he was immediately rowed to the other ship in the harbor.

The ship at anchor, originally called the *Mocha*, had been seized from the East India Company in a mutiny led by Culliford and Ralph Stout and renamed the *Resolution*. The captain had been killed in the uprising, and now Culliford and his men all had a price on their heads.

Within months the holds of the *Resolution* were filled with goods and precious metal. The pirates had robbed so successfully that Culliford began to fear retribution. He was shorthanded as well and decided to take his ship off the high seas so he and his men could rest and divide up their booty. He ordered a vacation on the island of Sainte-Marie.

LIKE HIS MEN, CULLIFORD thought Kidd was there to arrest them. Kidd quickly put his mind at ease. According to witnesses, he said to Culliford, "I would rather my soul broil in hell than I do you any harm."

Culliford was delighted to hear such a proclamation from his former captain and granted Kidd's hopelessly leaky ship safe harbor. With some towing help from longboats, Kidd nosed the *Adventure Galley* as close to the shore of the careening area at Îlot Madame as her deep draught would allow.

There she could stay until the rest of the crew arrived on the *Quedah Merchant*. With the entire crew to help, they could pump the water out and pull the ship onto the sandy shore to caulk and repair the damaged hull. Or so Kidd thought.

As a sign of friendship, or perhaps to lessen Culliford's suspicions,

Kidd presented Culliford with four cannons to bolster his ship's fire-power, and some shirt cloth. Culliford was very grateful for the gifts. As a return gesture he presented Kidd with about five hundred pounds. Then he ordered to the quarterdeck jugs of *bumbo*, which quickly brought levity to the proceedings.

Once again Kidd assured Culliford that he had no intention of arresting him or any other pirates on Sainte-Marie. Capturing a pirate ship like the *Resolution* and a captain of Culliford's stature might well have solved all of Kidd's problems back home. Kidd knew this. But there was no way he could do it. After months of his despotic command, his crew hated him, and for the time being, Kidd simply felt lucky to be alive.

Against the setting sun the two crews mingled and drank *bumbo*. They told stories of voyages past and treasures to come. They were part of a fraternity, these "brethren of the sea," these pirates of the Indian Ocean.

PART II

THE SECOND EXPEDITION

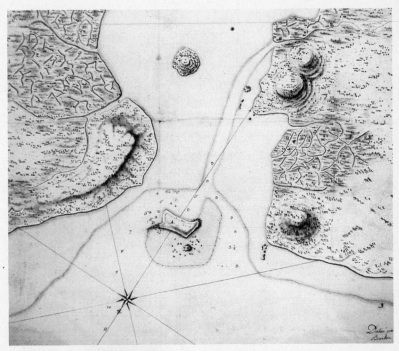

*There never was a greater
liar or thief in the world
than this Kidd.*

—EARL OF BELLOMONT

13

RETURN TO
TREASURE ISLAND

"YOU MUST RELAX," SAID GILLES GAUTIER, a tense smile glued on his face as we walked through the Tana airport at midnight.

"How can I do that?" I asked. "We just got double-crossed by the government."

We had returned to Madagascar on May 23, 2000, after four months back in the United States. During our time away, we had worked hard to make sure that we would not have permit problems. Along with the producers at the Discovery Channel, we'd spent considerable time and energy explaining the purpose of our expedition to the appropriate people at Madagascar's Washington embassy. Based on our information, we were led to believe that permits to film and excavate would be issued to us when we arrived in Madagascar. Now Gautier was delivering bad news: our permit had not been issued. We could not touch an artifact or even get into the water near the site. The site of the supposed *Adventure Galley* was completely off limits to us. Over the telephone Gautier declared that permission to excavate had been issued by the minister of the interior. When he asked to have it in writing, he had been refused. At that point we were in the air and there was no way to reach us with the bad news.

FRENCH GUIDE GILLES GAUTIER, WHO INSISTED ON DOING THINGS THE MALAGASY WAY. *Paul Perry*

"No, no, you have not been double-crossed," declared Gautier, his voice rising in mock indignation. "This is the way it is here. You have some of your permits, that's for sure. But some are yet to come. Just relax. You are on Malagasy time now."

The thought of being on Malagasy time frightened me. The clock was ticking. The Discovery Channel had budgeted enough money for three weeks of hard excavation work, and absolutely nothing for sitting around waiting for permits. I had a team of divers and archaeologists on the plane with me and a team of remote-imaging technicians scheduled to follow in two or three days. With them would come a camera team from Discovery, headed once again by David Conover. None of us had any room in our budgets for downtime. Now we were being forced to wait for permits that we already thought we had.

"I don't get it," I said to Gautier, trying to calm myself. "The ministers agreed to our petition and gave us permits saying we could dig. Everything was in order."

"That is true," said Gautier, as we stood in the customs line while armed officials slowly examined our luggage and then selected a few cases they wanted us to open for further inspection. "But the paperwork got forwarded to the Washington ambassador, and he would not issue the permits. He overruled the ministers here in the country, because he felt that the conditions were 'vague.'"

"Vague?" I asked. "What does he mean by vague?"

Gautier shrugged and then smiled like he always did. "I don't understand myself," he said. "None of the letters of authorization from the ministers seemed vague to me. Perhaps he just doesn't like you, Barry."

I thought about that for a moment and went on through the airport doors and out into the damp night of Tana. The gauntlet of service people lining the sidewalk brought my attention back to the immediate problems behind me, hundreds of pounds of exploration and dive equipment being piled on carts. Cabdrivers tapped our arms and asked if we wanted a ride. Baggage handlers made eye contact and tried to wrangle our bags from us. Gautier waved them away, but they came back like hungry flies and followed us all the way to the giant bus that had the words LES LÉZARDS DE TANA written across its sides.

Gautier pointed to a few hearty baggage handlers and told them to unload the carts into the bus. When they were finished, he handed them a few francs and closed the baggage door.

I must have looked worried because he shook my arm and offered an impudent grin. "Do not worry," he said. "It might take time, but I have good connections in this country. You will get what you want, or almost."

I winced. In a country like this one, "good connections" usually means bribes. I took a deep breath of the night air and found that it didn't smell so good. Then I got in line and followed my crew into the bus for the long trip to town.

WE ARRIVED AT THE Hotel Bonsoir well after midnight. The small lobby was dark, and the half-asleep night clerk was slow at processing whatever paperwork was required to get us into the rooms.

Most of the luggage, especially the heavy stuff, was left in the bus, which was then driven to a fenced compound guarded by dogs and patrolled by armed men working for Gautier. Theft was rampant in the capital city, said Gautier, but almost nonexistent on Sainte-Marie, where we would be headed in the afternoon. "Except for pocket-knives," he declared. "They can't resist stealing pocketknives."

Personal luggage and carry-on bags were wrestled up the stairs by the expedition members as they made their way to their rooms. In the next few days the crew would become much larger. In addition to the camera crew of four, several acoustical engineers were joining us from Witten Technologies, a company with equipment that could "see" underground with sound waves, producing three-dimensional images of objects buried beneath the earth. We had planned to use them to explore both the wreck site and the tunnels of Pirate Island.

But now as I sat in the lobby I wondered if either the camera crew or the technology guys would be able to come. Without permits to at least film the site and test it with the nonintrusive electronics equipment, Discovery Channel would cancel their trip and probably our expedition. It was a dark night.

In the morning I got my second wind. I pounded on John de Bry's door and told him to meet me down in the lobby café. By the time he arrived I was already halfway through a plate of fried eggs. My night had been plagued by Lariam dreams, wild fantasies caused by the anti-malaria medication that we had to take to keep from getting one of the world's most deadly diseases.

De Bry came downstairs and sat with me.

"What do we do now, chief?" he said.

"We get the permits," I said. "I don't know how, but we have to get the permits."

Our prospects didn't seem bright. It was Sunday, a day when no official business with the government would get done. Gautier assured us of this fact by repeating again and again, "These things take time." He would mutter those words whenever we wondered out loud about the possibility of expediting the mysterious permit process.

With the help of Gautier, we developed a plan. Since de Bry spoke fluent French, he would stay in the capital city of Tana with Gautier while the rest of us went to the island to prepare for whatever kind of permit we might receive. It was my plan to have all of the equipment ready and everyone in top shape so we could begin as soon as the permits were released.

"Being a little optimistic, aren't you?" asked de Bry when I told him what I wanted to do.

"Yes, Barry," said Gautier. "These things take time."

"I know that," I said, motioning to the members of the crew that were eating breakfast at neighboring tables. "But I don't want these guys to just sit around this city doing nothing. We'll have trouble then for sure. At least they can spend their energy on the island in case we do get going."

"You must relax, Barry," intoned Gautier. "These things take time."

"Yes, but I hope not too much time," I said. "We don't have it."

With that, de Bry was left behind with Fabrice Digiovanni, Gautier's partner, whose unpleasant task it was to lead our archaeologist through the maze of bureaucracy toward the prized permits. Digiovanni didn't look happy about his task. With his black hair swept back in a ducktail and a French cigarette fuming at his lips, he frowned as I talked to de Bry about pressing the urgency of our situation with the cabinet ministers.

"There is no fast here, Barry," said Digiovanni. "Everything here takes a lot of time."

I glared at Digiovanni and then Gautier. Both of them just shrugged.

"Good luck, John," I said to de Bry as Gautier and I began our trip to Sainte-Marie.

"I'll need it," he replied.

THE SOURCE OF OUR permit problems became clear when de Bry spoke to the minister of transportation, an affable bureaucrat who had no problem personally with the notion of issuing a permit for us to dig. In fact, he said, none of the cabinet ministers had any opposition to us searching for Captain Kidd's ship. The opposition, he said, was coming from the Washington ambassador.

"He keeps saying he needs more information," the minister told de Bry, who thought that was some kind of code requesting a bribe. But when he asked the cabinet minister if the ambassador required "special permit money," he received a negative response.

"This isn't about money," the minister said. "This is about something else. I think someone is trying to stop you."

I was puzzled when de Bry called me. *Who would try to stop us?* I wondered.

"It sounds like it's another archaeologist," said de Bry.

"Why do you think that?" I asked.

"It sounds to me like someone has been talking to the ambassador," said de Bry. "The minister here told me that suddenly the ambassador became very cold toward your expedition. Everything seemed to be 'go,' and then suddenly it was 'stop.' Sounds to me like someone torpedoed us."

In my head I ran through a list of people who might do such a thing but eliminated them one by one. I could think of several people who didn't like me, including some archaeologists who are against private archaeology projects like my own. But the notion that a legitimate archaeologist would stick his or her neck out by lying about me seemed far-fetched. Or did it?

"Who do you think it is?" I asked de Bry.

He paused for a moment. "It might be Dick Swete," he said.

"Who?" I asked.

"You remember. That guy from Sacramento who was on the island looking for the *Serapis,* the ship that John Paul Jones captured from the British."

I thought for a moment but drew a blank. "Did I meet him?" I asked.

"No, none of us did. He had left the island a few weeks before we came," said de Bry. "We heard about him when we were trying to track down unbroken pieces of porcelain from the wreck . . ."

As de Bry recounted the story, I began to remember what had happened. After bringing up some broken porcelain cups from the wreck site, de Bry had the idea of asking the islanders if they had any old china that might be unbroken. Not only would some whole pieces of china give him a complete picture of the porcelain on the site, it might lead to more artifacts, ones that had been handed down through generations.

Soon word came back that the owner of a nearby hotel had a nearly complete platter of blue porcelain that an archaeologist had found only weeks earlier. De Bry and Paul Perry went to the hotel to see it. As it turned out, the platter was thick and looked nothing like the fine porcelain we were working with.

The hotel owner told them that an archaeologist named Dick Swete had visited the island several times over the past few years looking for a ship named *Serapis.* De Bry knew immediately that the ship had been taken by John Paul Jones in his epic battle against the British in 1779. Jones, fighting from the deck of his *Bonhomme Richard,* had his crew nearly decimated by a broadside of cannon fire from the *Serapis.* When the English captain asked Jones to surrender, the American captain shouted, "I have not yet begun to fight!"

Jones went on to defeat the British ship at the cost of his own. As the *Bonhomme Richard* sank, Jones and his crew transferred to the *Serapis.* De Bry knew nothing of its history after that.

The hotel owner said that Swete and a few divers had found the wreck site only a few weeks before we had arrived for our first expedition. He showed de Bry and Perry a nautical chart of the island and pointed to a spot well outside the harbor where the *Serapis* had been found.

"Swete became so excited when he found the ship that his mouthpiece fell out and he nearly choked to death on seawater," said the hotel owner. "He found this china down there and also the anchor that is in the lobby. He gave that to me as a gift."

De Bry wrote down Swete's name and phone number so he could call him once we got back to the United States. He wanted to make contact with him as a professional consideration, so he would know that we were on Sainte-Marie searching for the *Adventure Galley* and not trying to encroach on the *Serapis*.

De Bry and Perry had told me about this during the first expedition, but I had been so focused on finding the *Adventure Galley* that I had forgotten.

"What makes you think he's working against us?" I asked.

"Nothing, really, just a hunch," said de Bry. "I called him after the last expedition and told him that we were working on a shipwreck close to his and that we had no intention of disturbing his excavation. It was a courtesy call, that's all."

"What did he say?" I asked.

"He didn't seem very happy to have company on the island," said de Bry. "He was grouchy when I told him you planned to go back to the island. When I told him that we were working on a Discovery documentary that seemed to bother him even more."

"Did he say why?" I asked.

"No. When I told him he muttered, 'Oh, shit' under his breath."

I could see why the presence of a film crew would disturb him. It would be upsetting for me to see other television projects working close to one of my wreck sites.

"What happened then?" I asked.

"Not much. He thanked me for calling, said he was busy, and that was pretty much the end of the conversation," said de Bry. "He did say that he had dealt with you before. He didn't say where, but he didn't seem happy to be dealing with you again."

That last sentence stopped me. I rummaged through my mind in search of a face to put with the name Dick Swete, but I couldn't produce one. I asked de Bry if there was anything distinguishing about Swete, something that might trigger my memory.

"He has a prosthetic leg," said de Bry, who had asked other archaeologists about Swete. "And he's very skinny. That's all I can tell you about him."

I hung up the telephone and tried to place Swete. I thought about every shipwreck I had ever worked on and all of the characters associated with them. I couldn't think of anyone who fit the description of Dick Swete.

I went to the gear locker where Wes Spiegel was arranging the dive equipment and asked him if he had ever heard of Dick Swete. If anyone would remember, it would be Spiegel, since he had worked with me for almost fifteen years. Spiegel just shook his head when I mentioned the name. "Doesn't ring a bell," he said.

That was how I left it. I didn't remember Dick Swete, and I really didn't think he was trying to stop our expedition.

But I was wrong, very wrong.

To keep morale high, I told the crew to do anything they wanted for recreation, as long as they didn't get hurt. Some went surfing, while others went hiking or diving. I spent most of the time at the hotel working the telephones and hoping for good news. We had switched to a new hotel a week into this expedition, the Princesse Bora, after one last meal of green chicken and dirty rice at the Orchidées Bungalows made a number of crew members sick. The Bora was owned and operated by Fifou Mayer, a Swiss born in Madagascar who had a firm grasp of its laws and customs.

TOP: EXPEDITION MEDIC TODD MURPHY TENDS JENNY CLIFFORD'S SORE THROAT. *BOTTOM:* JENNY SKETCHES THE WRECK SITE, WITH THE CAUSEWAY IN THE FOREGROUND AND THE OLD CATHOLIC CHURCH ACROSS THE BAY.

Barry Clifford/Paul Perry

"You should ignore what the Washington ambassador says and dig anyway," he told us. "He doesn't have any say in this country, and you already have the backing of the minister of transportation."

I disagreed. The producers at the Discovery Channel wanted to stay in good graces with the Washington ambassador in case they needed his help with any other programs. What they said would go.

"Until this gets resolved, I guess we're having a tense vacation," I told the crew during a celebration for my fifty-fourth birthday. Fifou, an accomplished chef, had made a beautiful chocolate cake, and I blew out the candles as the crew sang "Happy Birthday."

"What did you wish for?" asked de Bry, who had spoken to everyone he could and was now on the island awaiting the outcome.

"You can probably guess," I replied.

After fourteen days of waiting, the Washington ambassador finally caved in and allowed us to get to work.

News of the permits came via a late-night telephone call that I made to David Conover, the Discovery Channel producer who had joined us on the first trip and was now planning a return visit. At the time I didn't know why the ambassador finally acceded to the pressure being put on him by the Malagasy cabinet ministers as well as the Discovery Channel. All I knew was that we had the permits and could now begin our work on the site.

"There's just one hang-up," said Conover, his voice sounding like it was coming from a tin can as it bounced off a satellite and into my receiver. "The permits won't allow us to actually dig on the site. They only allow us to photograph what's there on the surface. If we dig into the ballast mound we run the risk of getting kicked out of the country."

I couldn't believe what I was hearing. Essentially that left us with what one member of the expedition called a "window-shopping permit," one that allowed us to look but not touch.

Ordinarily such a limited permit would have killed a shipwreck-hunting expedition. After all, what good is a permit to explore an archaeological site if it doesn't allow you to dig? But the Washington ambassador didn't know about our secret weapon. The acoustical equipment and engineers coming from Witten Technologies would allow us to see underground *without* turning a single spade full of dirt, on land or sea. It was like having X-ray vision.

Still, such a limited permit was not what I had in mind. I had hoped to be able to dig for those objects that were revealed by Witten's equipment. The point of bringing this gear in the first place was to find buried objects quickly so we could dig for them with as little impact on the site as possible. That would not happen, at least not on this expedition. We would be stuck with images of artifacts underground, objects we could only look at but not pursue.

"This is going to be like looking at an airport X ray of a container full of museum artifacts," I said to Conover.

"Exactly," he said. "Look but don't touch."

It was one in the morning when I finally got off the satellite phone with news of the permits. Everyone was sound asleep. Bob Paine and Spiegel had spent the day testing and cleaning equipment, while Perry and de Bry tramped through a swamp searching for a secret passage to the ocean that the natives said was used by the pirates. Brandon and Jeff Denholm had been surfing over the island's dangerously sharp coral reef, and Dr. Cliff Cloonan, a Special Forces medical doctor who joined us for the adventure, had spent the day with Ben Perry, Paul's sixteen-year-old son, treating natives for minor infections and malaria. Archaeologist Cathrine Harker and her husband, diver Chris Macort, had gone off alone and spent the day on a sunny beach at the tip of the island. At dinner their pigment-poor skin was beet red from overexposure to the sun.

And Gilles Gautier, of course, had occupied himself leisurely smoking while he pontificated on the unrelenting snail's pace of the Malagasy bureaucracy.

All of these activities had left everyone exhausted but me. I was energized. *The permits were ours,* or at least a limited version of the same. *Now we can begin,* I told myself.

Turning on a flashlight, I went from bungalow to bungalow, telling the crew that the permits had arrived. Some sat up with a start when I shined the light in their face to announce the good news. Others barely stirred. In my enthusiasm I wandered into a bungalow that was occupied by a couple of French tourists who had nothing whatsoever to do with our expedition.

"We have the permits!" I said, thinking it was Chris and Cathrine.

"Excusez-moi?" said the man, looking at me like a raccoon caught in a hunter's light.

"Wrong bungalow," I said, rushing outside.

I switched off the light and disappeared into the darkness. By morning, I was sure, the Frenchman would dismiss the event as a bad dream.

14

"WICKEDNESS SO GREAT"

THE *QUEDAGH MERCHANT* LIMPED INTO Sainte-Marie about five weeks after Kidd's arrival. Her deck was a mess of ripped sails and uncoiled rope, and her sailors had the look of having survived a near-death experience. Bollen was in no way the seaman that Kidd was, and it showed in the battered condition of the ship he'd commanded. High seas had buffeted the fully loaded vessel, and Bollen's poor navigational abilities had put them far off course. The *Quedagh Merchant* was brought into a spot next to the *Adventure Galley* and the *November*, and when Kidd saw the damage on the newly arrived ship he was not happy. But that was the least of his worries. Now safe on land and out of the clutches of their tyrannical captain, many of the sailors under Kidd's command let it be known that they had no intention of ever sailing with him again.

With the arrival of the *Quedagh Merchant* and the thirty sailors on it, the din against Kidd grew even louder. His crewmen gathered into an angry mob and demanded that the plunder taken from the voyage be divided. At first Kidd didn't want to divide the booty, as his agreement with Lord Bellomont and his other partners specified that the cargo not be broken up until the ship was safely back in Boston. This comment must have brought guffaws from the crew, since Kidd had already broken up much of the cargo when he sold booty in India from the *Quedagh Merchant* and the *November*. Realizing the futility of his argument, Kidd gave in. He ordered that the cargo from all of the ships be unloaded and spread out on shore.

A ramp for off-loading was rigged on the railing of each ship, and the men began lugging cargo out of the holds and onto dry land. The scene resembled a busy anthill as the sailors carried bales of cloth with striped or flowered patterns, bags of sugar and saltpeter, large ingots of iron and smaller sacks of powder, pistol shot, guns, even a couple of anchors. By the time they were finished, the careening area looked like a large general store that had lost its roof in a hurricane.

Along with all of these goods, Kidd brought out the money he'd made from the sale of much of the plunder in India. He had about ten thousand pounds total in gold and silver. By Kidd's account there were 115 crewmen left with whom the booty would be shared. Along with his quartermaster, Kidd divided the booty into shares, each one equaling as much as four bales of cloth, ninety pounds' worth of coins, and various smaller items. There was also a large amount of gold dust, silver and gold ingots, and precious and semiprecious stones; these were divided, too. Kidd then lay claim to forty shares, his portion of the plunder under the contract that he and all the men had signed. That supposedly came to 1,111 ounces of gold, 2,353 ounces of silver, and 69 stones, although there is strong reason to believe Kidd may have had far more booty not officially accounted for.

Each man was given his due according to the contract. Most of the able-bodied seamen received a full share of the goods and a half share of the money. The officers received more. After more than a year of scouring the high seas in search of plunder, the men of the *Adventure Galley* finally had their payday.

There are various versions of what happened next. Kidd claimed at his trials a few years later that he tried to convince his crew that they should now capture Culliford and his gang and return them to justice in England. If they did this, they would be living to the letter of their contract with the king and the trespasses they had committed by robbing friendly ships might be overlooked.

The mutinous crew would have none of this, said Kidd. When he implored the men to join him in capturing Culliford, one of them shouted: "We would rather shoot two guns into you than one into the other!"

"Make that four guns!" declared another. There was a roar of agreement as the men spewed forth a torrent of accusations about the captain.

By the end of the day, ninety-seven men left Kidd. They had been talking to Culliford's crew for weeks now and saw their future with the captain of the *Resolution*. The men left the scene en masse, picking up their bundles of booty and heading for Culliford's ship, which was anchored nearby.

Kidd was left with eighteen loyalists and three damaged ships.

He took to the *Adventure Galley* for his own protection. He later claimed that, through the grapevine, he had heard that the crew was going to return and seek further retribution. To protect himself, Kidd loaded forty pistols, which he laid out all over his cabin so they would be within easy reach. Then he used bales of cloth to barricade the cabin door.

For the next four or five days, Kidd claimed, his former crewmen raided the *Adventure Galley* relentlessly. According to his account they boarded their former flagship and "carried away great guns, powder, shot, small arms, sails, anchors, cables, surgeons' chests, and what else they pleased."

When they heard that Kidd had moved his share of the booty to Edward Welch's house for safe keeping, a number of them made the four-mile trek and demanded that the chest be opened. Welch did not stand in their way, and the mutinous crew broke it open and walked away with 10 ounces of gold, 40 pounds of plate, 370 pieces of eight, and Kidd's journal—the latter being a most convenient loss.

As the time came for Culliford's ship to sail, the anger of Kidd's crew boiled over for the last time. Emboldened by their imminent departure (and probably a healthy slug of rum), some of the men threatened to break in Kidd's cabin door and put an end to him. This was a night of terror for Kidd. Outside his locked cabin the men continued to threaten him, banging on the cabin's bulkhead and trying to push the door in. Still Kidd held fast. At one point he threatened to fire through the door. At another point he told them that he had many pistols "ready charged" and would kill several of them before they could put an end to him.

The men eventually tired of the confrontation and left. But not without doing considerable damage. Before the men boarded the *Resolution* and ventured into the Indian Ocean with Culliford, they stripped the *November* of anything worthwhile, including cannons, anchors, and even as many nails as they could pry from her wooden decks. They then burned the ship.

On June 15, 1698, the *Resolution* left port with about 130 men. Kidd and his meager crew watched as the stout little frigate sailed by. As it turned out, Culliford and crew would have an amazingly successful voyage, taking a rich pilgrim ship, the *Great Mahomet*, which scored them enough money to retire for life. Many of them did just that, returning to Sainte-Marie in December of that year to live the life of kings.

Had Kidd known the success Culliford was going to meet he might have tried to join him. Now, as the *Resolution* sailed past the *Adventure Galley* in the harbor of Sainte-Marie, Kidd stood at the rail of his flagship and scowled. His feelings about the men who left him was summed up in the narrative of the voyage that he wrote for Lord Bellomont. In the midst of describing their acts of plunder, Kidd wrote: "Their Wickedness was so great . . ."

WITH THE *RESOLUTION* GONE, Kidd's focus turned to his other problems. With no one to man the pumps, the *Adventure Galley* had filled with water and sunk. Now its keel was resting in the harbor mud, its deck slanted sternward. Had he been so inclined, Kidd could have put a cannonball on the forecastle and watched it roll the length of the ship and off the poop deck.

He would have liked to refloat his waterlogged flagship, but with only thirteen crewmen there was no way to bail her out and repair her hull. The Indian sailors who had previously manned the pumps had gone with Culliford, and Kidd could find no extra sailors who wanted to sail with him.

It was with a heavy heart that Kidd did what he did next. He ordered that everything of value be moved from the *Adventure Galley* to the *Quedagh Merchant*. Then, with his flagship empty, Kidd put the torch to it, burning his beloved vessel down to the waterline.

It's not hard to imagine what it was like for Kidd to burn his own ship, the once sleek hull of the *Adventure Galley* nosed up against the shore, perforated like Swiss cheese from worm damage. Kidd probably stood right underneath her on the beach below as the fire took hold. He certainly gazed in awe as the smoke became richer and darker and the flames turned a bright and ferocious orange. This was the end of his ship. But was it also the end of his dream? Did he now admit to himself that his mission had gone terribly awry? Did he now know that he was in terrible trouble back home? Did it seem as though his own life was going up in smoke?

To have commanded what was at that time the only ship in English history built specifically for pirate hunting must have filled Kidd with pride. And to transfer his association from that remarkable craft to the tubby Indian-built *Quedagh Merchant* must have felt like a comedown. He would be sailing that ship all the way to America. Teakwood construction and Indian design made it clear that it was stolen. A wide beam and deep draught made her slow and hard to maneuver. It would be an embarrassment, but it was all he had. It would have to do.

Throughout the night the *Adventure Galley* burned. In the morning, when smoke was no longer rising from the vessel, Kidd ordered his men to pound through the charred wood with hammers and find as many of the metal ship fittings as they could. By the time they finished, there was little left of the proud galley beyond the hull's bottom and a few ribs sticking above the water.

Like it or not, Kidd was now the captain of the *Quedagh Merchant*. She was shabby-looking, but Kidd had several months to repair her while he waited for favorable winds. Slowly and deliberately, he and his crew set about readying their new ship for the long voyage ahead. And slowly and deliberately, Kidd began to fashion stories and strategies to keep himself from the noose in his native land.

15

THE TECH TEAM

As soon as the permits were issued, more team members began flying in from the United States. Todd Murphy, one of my first divers on the *Whydah* excavation, came from Boston after being home only a few days from a business trip to the Middle East. He would function as the dive supervisor, a complex task requiring coordination, persistence, and above all, good humor.

The Discovery Channel film team, headed by David Conover, came on the same plane with Murphy. They were all jet-lagged but glad to be on the island. There had been a real chance that this expedition would be canceled, and the new arrivals felt a sense of unreality about being here, mainly because they never expected us to clear the permit hurdle this quickly.

A couple of days after the film crew arrived, a small plane landed at the tiny airport carrying sensitive electronic equipment and the five-man team designated to operate it. We called this "the flight of the nerds," but we didn't mean that in a derogatory way. There was no one we were happier to see than these techies. Without them, frankly, there would have been little we could do.

Heading the Witten Technologies team was Alan Witten, a geology professor from the University of Oklahoma and inventor of the "math-

ALAN WITTEN HOISTS GEM, HIS MATHEMATICAL LENS.
Paul Perry

ematical lens" that made the technology so amazing. He had created a complex mathematical algorithm that was able to read sound and radio waves, creating pictures of what was underground.

A type of subsurface imaging, this technology had been used by utility companies and city planners from Newark to Paris to locate such objects as gas mains and electrical conduits so roadwork could be done more safely and efficiently.

Although construction and civil-engineering applications were the bread and butter of his company, Dr. Witten's true love was using this acoustical brand of X-ray vision to explore the world of buried history.

In 1987, for example, Witten's technology located the remains of a partially exposed dinosaur skeleton that had been found by hikers. The twenty-foot skeleton, of a seismosauraus, was the longest ever found.

In 1992 Witten joined biblical archaeologists in Israel to excavate a prehistoric site in the Negev Desert known as Shiqmin. Using sound waves, they discovered the oldest complex of tunnels yet to be found in the Holy Land. So realistic were the images that the scientists were able to tour the complex on their computer screens. "We could literally fly through the tunnels like we were inside them," said Witten. "One time we got lost for a moment and panicked. It was as though we were really inside."

There was one theater that Witten's technology had not been tested in, that of undersea archaeology. Although the sound imaging works similarly to the way dolphins "see" in the water, it had never been used underwater outside of the laboratory. The attempt to capture a three-dimensional image of the *Adventure Galley*'s remains would be the first use ever of this technology in marine archaeology.

As a joke, Paul Perry and John de Bry picked up Witten at the airport in a zebu-pulled cart. He did not seem concerned that a zebu would provide their first transportation. Even after de Bry tried to rattle him by saying that the jolting trip to the hotel would take several hours, Witten simply shrugged and sat back on the cart's bench. Clearly he was ready for whatever challenge this assignment would hand to him.

Warren Getler, his director of communications, was another story altogether. From the moment he stepped off the plane, Getler worried about everything that a traveler to a strange country can worry about.

"I hear there are mosquitoes here that are Lariam-resistant," he said to Murphy, a Special Forces medic with deep knowledge of tropical diseases.

"Possibly," said Murphy, sensing an opportunity for some fun.

"And you have to be really careful about tuberculosis here, too, right?" asked the wide-eyed executive. "I hear it's making a comeback."

"That's probably true, too," said Murphy.

"And the food's kind of questionable here, isn't it?" he asked Bob Paine.

"Only the green chicken," said the fearless eater.

On the other hand there was Jakob Haldorsen, a senior scientist from Norway who feared nothing.

"What will you eat and drink if you are afraid?" he reasoned. "If I was afraid, I would never leave Boston."

Topping off this team were a senior scientist named Doug Miller and I. J. Won, a Mongolian-born scientist who owned Geophex, a company that specialized in finding land mines for the U.S. government.

With the Witten team on the island, things moved forward very rapidly. A huge geodesic-domed tent was set up at the wreck site and packed with equipment. On one side of the tent was a pile of dive gear bursting out of nylon bags. On the other side were two wooden tables loaded with computers and sound-recording equipment. Underneath the table were stacks of processing units to collect data. And all of it was connected to the outside world by a thick cable that snaked out of the tent and lay like a pile of spaghetti on the ground.

At the other end of the cable was the sensor head, also known as "the Fish." A heavy, torpedo-shaped object that emitted high-frequency sound from an underwater transmitter, it collected that sound through microphones and fed the waves up the cable to the Witten software where it would be massaged by the scientists in the tent to create a three-dimensional image of what lay below the ocean floor.

It took two days to set up the technology equipment. In addition to the above-water effort that was carried on by the Witten team, the underwater team strung rows of thin line that would connect to the Fish and guide it in a grid pattern back and forth across the ballast mound to obtain images of what was underneath it.

When it came time to collect data, the Fish was moved across the wreck site by a hand crank. It was slow work, but as it moved, a stream of data poured into the tent and showed up on the computer screen. When Haldorsen and Miller manipulated the data stream, a black-and-white drawing of the ocean floor began to emerge. One thin line at a time, an image of the suspected *Adventure Galley* was coming into view—after three centuries.

I didn't doubt that the final results would be exciting, if not spectacular, but the process of collecting the data wasn't. I began to think about some other places I could go, places I could explore right now. I settled on Pirate Island.

This island was perfect for Witten's imaging equipment. When we were here earlier, André Mabily had shown us several shafts that led to what appeared to be collapsed tunnels. Our brief visit had left us wondering if tunnels really did lace this island. Were they really receptacles for pirate gold? The quickest way to find out would be to use Alan Witten's X-ray vision.

"Someone go find André Mabily," I said. "It's time to go back to Pirate Island."

In Boston I had explained to Witten that one of the goals of our expedition was to examine Pirate Island for a supposed network of tunnels. The idea intrigued Witten, who rummaged through a pile of equipment and pulled out a long plastic case.

"This will give us plenty of insight," he said.

I didn't know what it was, but I knew that it would help solve the mystery of Pirate Island once and for all.

16

ENGLAND'S MOST WANTED

WHILE HIS CREW READIED THE *QUEDAGH Merchant* for the long trip to the Caribbean, Kidd pondered his fate. For a while he thought about staying in Sainte-Marie and establishing himself as a trader. He sold some of his booty from the *Quedagh Merchant* for gold and silver and may have thought about working with merchants in New York as an Indian Ocean fence like his old friend Frederick Philipse.

But as time wore on, it was clear that Kidd wanted to go home. Despite the danger he faced from the English legal system, Kidd somehow felt certain that he could talk his way out of trouble, convincing Bellomont and others that he had truly not become a pirate after all. Kidd may have thought that he'd taken enough plunder to satisfy his investors. Given that they were among the most powerful men in England, it might have seemed reasonable to the Scottish captain that they would protect him; particularly if he took care not to turn over all the loot to them at once.

In any case, the prospect of spending the rest of his days on this tiny island was not one that appealed to Kidd. Given the evidence at hand, it seems likely that Kidd loved his wife and wanted to return to her— not to mention that hiding out so far from civilization and society was a galling admission of failure and guilt.

Kidd tried to persuade as many white island residents as he could to join him. More than a dozen finally signed on. One member of his crew was Captain John Kelley, alias James Gilliam, a notorious pirate who was later hanged for his crimes in 1700. Another pirate, Edward Davis, was also on board. Davis, "an extraordinarily stout man," had served on the *Fidelia* under the pirate captain Tempest Rogers. He had been involved in piracy in the Caribbean and was arrested and jailed in Virginia a decade earlier. He and his fellow buccaneers won a long court battle and were even able to recover most of their booty (three hundred pounds was deducted from their plunder to start the College of William and Mary, which exists to this day). By the time Davis had reached Sainte-Marie with Rogers, he'd had enough of the pirate's life. He decided to hitch a ride back with Kidd and start again in a different profession.

Having hired some of the very men he was commissioned to seize, Kidd was ready to hoist anchor. The *Quedagh Merchant* was loaded with a rich cache of cloth, general merchandise, gold, silver, and jewels. Kidd even purchased slaves; not only to round out his crew but also as personal servants. Now he ordered his men aboard and raised sail to catch the monsoon winds that would carry him southwest, around the Cape of Good Hope and across the Atlantic Ocean to the Caribbean.

Though the court proceedings of Kidd's trial describe the trip to the West Indies as a "quiet voyage," some historians say that there is evidence that the voyage was anything but quiet.

Ken Kinkor found possible references to the murderous takeover of several ships by Kidd after he left Sainte-Marie. A letter from Lord Bellomont to the Lords of Trade written April 23, 1700, reads:

> Rear Admiral Benbow tells me that Kidd was so wicked as to murder all the Moors he took in the ships he made prizes of, in cold blood: and that he murdered several English and Dutch among 'em; only there were 10 or 12 young Moorish boys he saved, intending to make slaves of 'em, and one of 'em has some way or other got to Jamaica who has discovered this villainy of Kidd's.

There were also rumors of yet another mutiny having occurred on this voyage, one that Kidd reportedly put down at a cost of nearly forty lives.

Whatever may have really transpired, Kidd certainly had plenty of time to think of alibis for his behavior. He had a lot to account for when he finally faced Lord Bellomont and the other partners who had funded his Indian Ocean ventures. Not only did it seem obvious that he had turned pirate, Kidd had also violated most other articles of the contract, by stopping along the way and dividing and selling the booty.

How could he explain these violations of the agreement to Bellomont? And how could he explain robbing ships that were "friendly to the Crown," the most horrible offense of all? The French passes they had given him might lessen some of the violations. Even though he had flown a French flag and pretended to be a French ship to fool them into producing the foreign passes, these ships had actually declared themselves to be under the protection of the king of France.

But how about the ones that did not produce French passes, or that belonged to countries that were clearly friendly to the king, like the ship he had commandeered and was sailing on right now? The *Quedagh Merchant* could be difficult to explain to Lord Bellomont, as could several of the other ships he had taken.

Then there was the murder of William Moore to account for. *Damn my anger anyway,* Kidd thought. *I just wanted to straighten him out. I didn't think I would hit him hard enough to kill him. And with a bucket at that!* Although sea captains had great latitude when it came to punishment on the high seas, murder clearly crossed the line. Killing a seaman in the fashion that Kidd had done qualified as a capital crime. *Unless, that is, the man had been guilty of leading a mutiny . . .*

As Kidd paced the deck for those many months, he formed in his mind what he thought would be the perfect defense. He would insist to Lord Bellomont that the majority of the men on the *Adventure Galley* had decided to mutiny and turn pirate. He would say that Moore was one of the ringleaders. Although Kidd had unintentionally killed

Moore with his show of force, Moore's mutinous plans had spread like a cancer among the crew. Ultimately Kidd was unable to control them as they turned the *Adventure Galley* into a pirate ship.

IT IS CLEAR THAT Kidd finely honed this argument as he crossed the Atlantic. The letters that he wrote to Lord Bellomont once he arrived in the Western Hemisphere and the testimony he provided to court officials in Massachusetts and England show a man who believed he was the victim and not the perpetrator.

There was a large body of evidence, however, to refute his theory.

The files being compiled on Kidd's piracies had been steadily growing, but when he took the *Quedagh Merchant,* they suddenly mushroomed. As it turned out, the *Quedagh Merchant* had been leased to Muklis Khan, a good friend of the great mogul. The friend's fortune had been greatly reduced by the theft of this ship, and he loudly demanded that something be done about piracy in the Indian Ocean. His requests went straight to the one man who could enforce them, King William.

The great mogul was no fan of the East India Company. He saw it as a necessary evil, one that appeared to be a breeding ground for pirates. Pirates frequently learned their trade by first serving as sailors aboard company ships, and many of them returned to the Indian Ocean to pursue their line of work.

The mogul had demanded many times that the English government send squadrons of warships to his shores to get rid of the pirates that their country had bred. In 1695 an incident of piracy took place that almost put an end to the East India Company. Henry Avery, captain of the *Fancy,* took the *Ganj-i-Sawai* at the mouth of the Red Sea. The ship was filled with an immeasurable amount of treasure and many women, including, by some accounts, some of the mogul's daughters. On a windless sea, Avery and his crew lashed the vessels together and looted the Indian ship. Then as added insult they raped and tortured the sixty women on board for several days before sailing away, never to be captured.

This event had enraged the great mogul and his country. When word of what happened swept Surat, crowds of angry Muslims had surrounded the East India Company's compound and rioted. Had it not been for the intervention of local authorities, company employees would certainly have been murdered.

The governor of the region, Ahmanat Khan, had ordered the arrest of sixty-four employees of the company, including its president. They were held in a stifling Indian prison for eleven months and would have stayed longer had it not been for the intervention of a well-connected friend of the company.

The great mogul and many others in India believed that all pirates were English. The other nationalities that he did business with—Dutch, Portuguese, French, Spanish, Danes, and others—encouraged this belief. Their insistence that Englishmen committed most of the acts of piracy in the Indian Ocean helped them evade responsibility for piracy committed by buccaneers from their own nations.

The English were sympathetic to the great mogul's demands that their government send its navy to fight piracy off his shores, but they were strapped. Their seemingly endless war with France was tying up all of their ships, leaving none available for police duty.

When the *Quedagh Merchant* was taken, the old wounds of English piracy were reopened. Riots broke out in the streets of Surat, and things became so bad for members of the East India Company that they abandoned their homes and moved behind the walls of their factories. Life became even worse when local merchants refused to sell food to Europeans as a protest against piracy.

The regional governor demanded that the company pay compensation for goods lost on the *Quedagh*. When Samuel Annesley, the company's president, refused, the mogul threatened to purge his country of all Europeans. Such a move would have had a profound economic impact on Europe and ruined any number of English fortunes. Still Annesley refused to make any payments.

Pressure from the Indian government increased with Annesley's refusal. It stopped English trade in its tracks, refusing to let cargo be loaded onto any ships until its demands were met. Finally the company's desperate president agreed to compensate some of the shippers who had goods on the *Quedagh Merchant*. He also agreed to provide convoy protection for ships from India that were sailing to the Spice Islands.

The East India Company had had enough of piracy. They decided to focus their ire on Kidd. Although he was not the only pirate on the high seas, he had certainly become one of the most visible. The company decided to use him as a symbol of the troubles they were trying to eliminate.

In November 1698 Annesley wrote to the Lords Justices telling them about Kidd's seizure of the *Quedagh Merchant*. They formally accused him of piracy and demanded that an expeditionary force be sent to apprehend him. By return mail they were informed that Captain Thomas Warren and a small force were sailing to "pursue and seize Kidd if he continues still in those parts."

A "circular letter" was also prepared by the Lords Justices for delivery to the governor of every American colony. This letter called upon each of them to "take particular care for apprehending the said Kidd and his accomplices wherever he shall arrive . . . as likewise to secure his ships and all the effect therein, it being Their Excellencies' intention that right be done to those who have been injured and robbed by the said Kidd, and that he and his associates be prosecuted with the utmost rigour of the law."

Perhaps it was the next document to be produced by the English government that was the final knot in the noose that would eventually encircle Kidd's neck. In 1698 King William issued a royal proclamation on piracy, offering a pardon to all the pirates at Madagascar. The proclamation read in part: "Now we, to the intent that such who have been guilty of any acts of piracy in those seas, may have notice of our most gracious intention, of extending our royal mercy to such of them as shall surrender themselves, and to cause the severest punishment according to law to be inflicted upon those who shall continue obstinate. . . ."

The pardon extended only to those surrendering to Captain Warren, whom the king had been "graciously pleased to empower." It also made another thing perfectly clear: the pardon did not extend to Captain William Kidd.

He was now England's most wanted man.

17

THE TUNNELS OF PIRATE ISLAND

WE NAVIGATED THE BAY TO PIRATE Island in a tippy fleet of dugout canoes that André Mabily had assembled from the local natives. When we reached the island's wooded shore, Alan Witten climbed out of the dugout and lugged the plastic case to the top of the island.

"Where are the shafts?" he asked, getting down to business.

Mabily and Gilles Gautier showed him the three shafts. Witten looked down into each one and then asked the two guides to cut a swath through the underbrush so he could walk easily between them. Machetes were produced, and with wide sweeps of the blades the underbrush was hacked away.

While the path was being cut, Witten unpacked what looked like a fluorescent light fixture with a shoulder strap. The device was called a GEM, a Geophex electromagnetic sensor. According to Witten, the GEM broadcasts radio waves into the earth, mapping its density to a depth of about fifty feet; when the data is processed, empty spaces like tunnels show up on the computer screen.

With the pathway cleared, Witten slung the GEM over his shoulder and walked between the two shafts, the unseen radio waves beaming

FRENCH GUIDE GILLES GAUTIER PREPARES TO EXPLORE A PIRATE ISLAND
TUNNEL. *Paul Perry*

into the ground. In a few minutes it was over, and once we returned to
the bungalows his technicians would be able to provide a picture of
what—if any—tunnels existed on Pirate Island.

I was impressed by how much information could be gathered in so
short a time. Archaeological sites are usually a work zone of people digging and hauling dirt. Searching for these tunnels the old-fashioned
way would have taken days and possibly even weeks. And you are essentially destroying a site when you excavate it. Witten's low-impact technology could usher in a new age of archaeology in which scientists
could see underground without the tremendous site trauma of tradi-
tional excavating.

"I'm impressed," I said to Witten as we prepared the dugout canoes
for the trip back to the shore.

"And I'm not," said Gautier, spewing a noxious cloud of cigarette smoke from his grinning lips. "We cut jungle all day, and it takes only a few minutes to look into the center of the earth? What kind of archaeology is that?"

"Gilles, shovels are passé," said Witten, balancing his way into the dugout canoe. "What did you expect? The days of dynamite and big booms should be over."

THAT NIGHT A BIG boom came. Witten called several of us to the porch of his bungalow, where Doug Miller and Jakob Haldorsen were tweaking an assembly of color on a laptop screen. The green and blue with a swath of orange were assigned by the technicians to show density. The green and blue were solid earth, while the orange represented vacant space.

"This picture indicates very strongly that there are tunnels on Pirate Island," said Witten. "See the orange? That is a feature consistent with a tunnel."

"What do you mean?" asked Wes Spiegel. "Is it a tunnel or isn't it?"

"My data tells me that it is an area of low conductivity, meaning it has more water or air in it than the surrounding area," said Witten. "Indications are that there is a tunnel there. But just because it looks and smells like a tunnel on the computer screen doesn't mean that it is really a tunnel."

Despite Witten's caution, excitement rippled through the crew. Not only had we found a graveyard of pirate ships and the likely site of the *Adventure Galley,* we had now almost certainly confirmed the local belief that tunnels laced Pirate Island and formed a likely underground fortress for the pirates who inhabited it.

"What do you think it was like on that island?" Jeff Denholm asked John de Bry.

"Pirates were smart," he said. "They probably dressed up in strange clothing to scare the natives, and this convinced them that the island was haunted."

"Maybe they even buried some of their fellow pirates there," I speculated. "Anything to convince the locals that this place was taboo."

If those are tunnels, Witten was asked, how are they configured?

Picking up a pen and paper, Witten drew a blueprint of a tunnel complex that dropped down two levels and spread out across the entire island. At one end he drew a tunnel that extended into the bay and had a gate across its mouth. "That is a shaft that leads into this lower chamber," said Witten. "If anyone came into the bay to capture them, they could open this gate and flood the lower tunnel."

"Why?" I asked.

"So no one could get the treasure," he said. "If this island was a pirate bank, then the lower chamber would be their vault."

While we were on Pirate Island, someone had drawn a large skull and crossbones on the side of the geodesic tent. Immediately the locals began referring to the structure as *tente des forbans*, French for "tent of the pirates."

The scientists working inside the tent looked like anything but pirates. In their button-down shirts and khaki pants, they looked more like engineers heading for work at Motorola or Intel than technicians on an expedition to examine a sunken pirate ship. They were working long and hard hours collecting data from the site in hopes of getting a clearer picture of the ship that had settled deep into the mud.

This was more difficult than it might seem. Because the Fish was being cranked by hand under the water, it moved in a jerky fashion and did not gather information in a smooth and consistent way. This made it impossible for the software to process the information correctly. Another problem was the line that the Fish moved on. It was clothesline, and it wiggled as the Fish was being cranked, and drooped as the heavy device crossed its middle point. This unanticipated movement added another element of doubt to the data being fed into the computers.

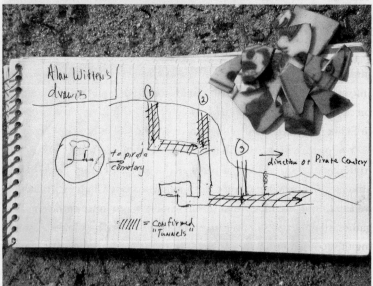

PIRATE ISLAND AND ALAN WITTEN'S DRAWING OF THE TUNNELS DUG
THROUGH IT.   *Paul Perry*

Miller and Haldorsen, who were accustomed to gathering data in a meticulous way with few variables and whose equipment had never before been used underwater, were becoming increasingly worried about coming up with any meaningful results at all.

The scientists continued to gather undersea data for several days, while Witten, his curiosity piqued by the possibility of more tunnels on Pirate Island, had more swaths cut through the underbrush.

"The tunnel that's the most interesting is the one that supposedly goes under the bay," said Witten. "It seems physically impossible because it would fill up with water as quickly as they dug it. But all the natives insist that there is one there." He scoured the island once more with his detector, then packed up his equipment for the trip home.

In the time that it would have taken archaeologist Howard Carter to assemble a shovel team at King Tut's tomb, the group from Witten Technologies was here and gone. They had collected several gigabytes of raw data and were on the airplane home to process it in their own laboratory in Massachusetts.

With the techies gone, it was somewhat difficult to keep the level of intensity up. We were still waiting for permits from the Ministry of Information and Culture that would allow us to remove just one small piece of wood from the wreck so it could be sent to a special laboratory in England for dating.

We invited the director of the minister's cabinet, Limby Maharavo, to the island to see the work we were doing. He was impressed by the amount of archaeology that could be done without digging, but when we asked him to authorize us to dig some test pits around the site and gather wood samples, he refused. Instead he asked for a letter explaining the goals and purpose of the test pits, which he promised to give to the minister of culture himself.

We wrote the letter and continued to wait. No response from the minister came.

The Discovery Channel filmmakers dove on the wreck and in other spots around the island, gathering "B roll," the footage necessary to fill

DOUG MILLER *(LEFT)* OF WITTEN TECHNOLOGIES AND JOHN DE BRY WATCH
THE CONTOUR OF THE WRECK SITE DEVELOP ON A COMPUTER SCREEN.
Chad Henning/Discovery Channel

in between pertinent action, but for the others there was little to do.
We had already examined just about everything we could and didn't
think we could find more without digging into the ballast mound or
around the hull.

What was making us most anxious was the fact that without the samples we wouldn't be able to establish which one of the several wrecks in
the harbor was the *Adventure Galley*. There were three other distinct
wreck sites, all of their locations corresponding with the 1722 account
in the English captain's logbook. And though we had great side-scan
sonar images of these ballast mounds, all were within three hundred
feet of one another. Without authorization to dig, there was no way to
identify Kidd's ship.

18

FALSE REDEMPTION

 PULLING INTO A PORT AFTER A LONG
sea voyage is usually a relief, but for Kidd
and the crew of the *Quedagh Merchant,* arriving on the Caribbean island paradise of
Anguilla in April 1699 was more of a shock.

They came to this lonely outpost hoping
to log several days of rest and relaxation.
Their stores of food and water were seriously depleted by the seemingly endless journey, and several of the crew were sick, including
Kidd's brother-in-law Samuel Bradley.

The white sand beaches of this English colony were warm and invit-
ing, and the small town offered a chance to drink alcohol and relax in
an environment that wasn't constantly pitching and rolling. As soon as
they set foot on solid ground, however, Kidd and his crew received
news that they were wanted by none other than the king of England.

Their shore leave turned into a hurried four hours, as Kidd quickly
purchased what food and water they could carry before rowing back to
the *Quedagh Merchant,* heading for Saint Thomas, a Danish port known
to openly accommodate pirate commerce.

Upon arrival he immediately paid a visit to Governor John Laurents
to ask for official protection against English ships. Despite repeated
requests, the governor refused to sanction his visit. He'd heard about
the letter from the Lords Justices ordering all English governors to

arrest Kidd on sight and did not want to side with a wanted man against such a powerful nation as England.

After two days in port, Kidd left. Before he did, though, he sent his brother-in-law ashore. The poor man had been a victim of "the bloody flux," a form of dysentery, for almost two years now, and Kidd hoped he could find some relief from one of the island's doctors. Along with young Bradley four other crewmen left as well, sailors who didn't want to be associated any longer with the infamous Captain Kidd.

Next stop was Hispaniola, the island that today is politically divided between Haiti and the Dominican Republic. On his way, Kidd was lucky enough to encounter Henry Bolton, a customs collector for the English government and a trader of questionable repute. When Kidd saw that Bolton was hauling a cargo of hogs, he immediately purchased twenty of them. As the squealing cargo was being transferred between ships, Kidd asked Bolton to find any buyers for the bales of cloth he had brought with him from the Indian Ocean.

Kidd brought the *Quedagh Merchant* into a river and moored her to trees on the riverbank. Within days Bolton returned with a fleet of small merchant ships hungry for Kidd's exotic wares. One by one the ships lined up next to the giant merchant vessel and off-loaded bales of cloth, with Kidd netting as much as sixteen thousand pounds from these transactions. Kidd's main plan now was to travel light and to leave the *Quedagh Merchant* behind.

Kidd and Bolton reached an agreement. Bolton and his twenty-two men were to stay with the *Quedagh* and sell the cloth remaining in her holds. Kidd purchased a sloop, the *San Antonio*, from Bolton and prepared her for the voyage to New England. He took only those men from his crew who wanted to come and left the others behind.

He was staking everything on a bluff. If he continued to run, he would certainly be captured and tried as a pirate, but if he returned voluntarily and pled his innocence, he might be spared.

Kidd caught the wind and headed north. Looking back at the *Quedagh Merchant* as he piloted the secondhand sloop toward the

Florida Straits, he didn't yet know that he would never see her again. Bolton stayed with the ship only a few weeks before abandoning it, taking what he could of the remaining bales of cloth and leaving the rest. With no one left to watch over her, the *Quedagh* was visited by islanders and curious sailors alike. Finally someone set fire to her worn deck. The ship burned to the waterline and then sank.

THE *SAN ANTONIO* HOPSCOTCHED her way up the eastern coast of what is now the United States the first week of June. They stopped at a town in the Delaware Bay with the colorful name of Whorekills. A sloop from New York City met them, and the crew spent two days transferring bales of cloth and chests filled with unknown goods.

Then Kidd pressed on.

The *San Antonio* sailed around the eastern end of Long Island to Oyster Pond Bay. Here Kidd wrote a letter to James Emott, a lawyer and friend, to bring his family and meet him at the bay. Kidd didn't have to wait long. Within a couple of days Emott arrived with Kidd's wife and two daughters.

The Kidds had not seen one another in nearly three years, but their joyous reunion must have been tinged with fear, with Kidd being the most wanted man in the colonies. By entering English territory governed by Lord Bellomont he was taking a great risk. On the high seas Kidd had become adept at dodging trouble, but here on land he would be easier prey. Virtually anyone who wanted could arrest or even shoot him.

With his wife and children by his side, Kidd told Emott about his alibi, claiming that he was not a pirate himself but a victim of pirates in his crew. In a letter written by Duncan Campbell, the postmaster of Boston, to Lord Bellomont, Kidd recounts how the pilot of the *Quedagh Merchant* presented himself as a Frenchman by producing a French pass, which Kidd could only conclude to mean that the ship was French and therefore fair game. "Whereupon . . . Kidd and Company took the said ship," the letter said.

Once Kidd realized that the ship's real captain was English, he told his men to release her. The men would have none of that, said Kidd. "[They] violently fell upon him, and thrust him into his Cabin, saying the said Ship was a fair Prize and then carried her into Madagascar and rifled her of what they pleased," the letter said.

When he further protested against the taking of the *Quedagh Merchant* and the *November* in Sainte-Marie, the majority of the crew rebelled. They stole guns and other things from the *Adventure Galley*, he told Campbell. Then, wrote Campbell wrongly, they "sett the said Galley on fire." What else was an honest sea captain to do but sail any available ship—even the *Quedagh Merchant*—back to his homeland, he asked.

Kidd's well-honed story painted him as a victim, a sea captain whose noble intentions had been subverted by the mutiny of an evil crew of pirates. It did leave out several essential elements, such as how he flew a French flag to trick the ship captains into thinking they were dealing with the French and how the crewmen forced Kidd into his cabin because they hated him, not—as he stated—because he wanted to return the ships.

He also failed to mention the murder of William Moore, though chances are good that he simply considered the incident insignificant.

He even presented physical evidence as proof of his story: the French passes he had received from the *Quedagh Merchant* and the ship they had renamed the *November.*

The people who first heard this story—his family and his lawyer—didn't know about the missing parts. They listened in awe as Kidd recounted his self-serving version of the three-year journey, an impressive story to be sure, spiced with self-importance and bravado. But was it convincing?

Emott, for one, was having difficulty believing his friend's tale. If the crew had truly turned pirate as Kidd claimed, why then was he allowed to return with so much booty? Emott held his tongue until Kidd announced his next step. He planned to travel to Boston and confront Lord Bellomont with "the facts," telling him the same story that he had just told them. Surely, said Kidd, this would clear his name.

Unconvinced, Emott offered an alternative plan, and Kidd, underneath the show of confidence, was just nervous enough to accept it. Write a letter instead with what you have just now told us, said Emott, and he would sail to Boston and deliver it to Lord Bellomont. That way he could test the waters and see if it was safe for Kidd to present himself in person.

Kidd's letter told of the crew's mutiny and piracy, of Kidd's futile attempts to honor his commission, of the loss of the *Adventure Galley* and the unavoidable need to sail the *Quedagh Merchant* to the West Indies. Then, with the bad news out of the way, Kidd tempted Lord Bellomont with a hefty prize. He was carrying ten thousand pounds of gold and silver with him, he said, and left behind in Hispaniola was an additional thirty thousand. If Lord Bellomont would pardon him and the few loyal crew members that remained, Kidd would sail back to the West Indies and retrieve the money.

Emott delivered the letter in a sloop commanded by Campbell. Kidd and his family sailed to Block Island off the coast of Connecticut to wait for Lord Bellomont's reply.

In short order, Emott returned. Carefully worded yet also encouraging, Lord Bellomont's letter said that he had never granted a pardon to anybody "without the King's express leave or command." However, he had listened to Kidd's story about renegade sailors having taken the ships, "violently against your will," and he acknowledged the receipt of the two French passes, evidence that the ships taken by his mutinous men had claimed to be French. Emott also mentioned—almost in passing—that Kidd had ten thousand pounds with him and another thirty thousand pounds "left in safe hands" in Hispaniola. Lord Bellomont wrote:

You showed a great sense of Honour and Justice in professing
with many asservations your settled and serious design all along to

do honor to your Commission and never to do the least thing contrary to your duty and allegiance to the King.

I have advised with His Majesty's Council and showed them this letter this afternoon, and they are of opinion that if yor case be so clear as you (or Mr. Emmot for you) have said, that you may safely come hither, and be equipped and fitted out to go and fetch the other Ship, and I make no manner of doubt but to obtain the King's pardon for you and those few men you have left, who I understand have been faithful to you and refused as well as you to dishonor the Commission you had from England.

I assure you on my word and on my honor I will performe nicely what I have now promised, tho' this I declare before hand that whatever treasure or goods you bring hither, I will not meddle with the least bit of them, but they shall be left with such trusty persons as the Council will advise until I receive orders from England here how they shall be disposed of. . . .

Your Humble Servant

Kidd breathed a sigh of relief. Lord Bellomont seemed to think that there was no doubt of a pardon. Kidd wrote a letter back, declaring his "most hearty thanks" for Lord Bellomont's understanding. "I do further declare and protest that I never did in the least act Contrary to the King's Commission nor to the Reputation of my honorable Owners, and doubt not but I shall be able to make my Innocency appear, or else I had no need to come to these parts of the world, if it were not for that, and my owners' interest."

Kidd felt relieved at the progress so far. He was certain that he would regain his good name soon. "Upon receiving your Lordship's letter I am making the best of my way for Boston," he wrote.

For Kidd, the trip to Boston would be a voyage of redemption, or so he thought. He had no reason to suspect any different.

19

PIRATE GOLD

 A FEW DAYS AFTER THE WITTEN TEAM left, a storm pounded the island. When it abated, John de Bry and I decided to go down to the site and see what we could make of the wreck.

As de Bry headed for the ballast mound, I swam a circle around the entire wreck site, looking for portions of wood that might be exposed. On my first loop I saw what appeared to be a cement block sticking out of the mud. Tapping it with my dive knife, I discovered that it was a coral-encrusted timber.

This would be a good place to retrieve a wood sample once the permit came through. When de Bry swam over, he was immediately drawn to a second timber near the one I had found. I had to head for the surface to get a new tank.

Sweeping his hand back and forth, de Bry began fanning the silt off the beam to see if there was any wood surface visible. Then he saw it: a gold glint that was no larger than the size of a fingernail. He fanned the tiny spot more, and it turned into a full-sized gold coin, glued to the beam.

He swept his hand again, and there was another coin, not far away from the first. Taking off his glove, he pried the coins off of the beam, shoving them down in the finger of the glove. He carefully marked the spot where he found the coins so we could return them later and surfaced.

After unstrapping his tanks and removing his weight belt, de Bry grabbed me by the elbow and walked me to a remote portion of the dock.

"What would you say if I found something that the pirates were almost certainly carrying that would prove whether or not this is Kidd's ship?" he said, opening his hand.

My jaw dropped. De Bry closed his hand quickly to conceal the coins, and we walked back to the tent, where I zipped the flap shut for extra privacy.

"What's that?" I asked, pointing to the writing on the coins.

"Arabic," said de Bry. "I know some of what is written on this coin because I remember it from the six years I spent working in Saudi Arabia. That word right there is *Allah*. This other coin is likely a Mogul coin from India, but I'm not sure. I can't see a date, so we'll have to photograph them and send the photos to experts in France. They can tell us when and where they were minted."

One of the two coins had holes through it, as though it had been strung and worn around a pirate's neck or sewed onto clothing like a button, a typical and unique pirate practice. We had seen similar coins on the *Whydah* site.

"Do you think these are from Kidd's ship?" I asked, hopeful.

"I don't know," he said, looking at the coins. "The *Adventure Galley* was filling with water and sinking by the time they got into harbor. There was no way that the sailors on board could get down below and recover everything, even if it was gold. They would have had to leave something behind, and since they had just finished robbing ships from India, it would make sense that the coins are an exotic variety such as these. But only the coin experts in France can tell us dates or at the very least the period."

I could hear a chatter among the crew as someone approached the

THE FIRST GOLD COINS AND WHERE THEY WERE FOUND,
ENCRUSTED TO A BEAM. *John de Bry*

tent. Gregory, our security guard at the site, began pounding on the
side of the tent as though it were a door.

"Mr. Barry! Mr. Barry!" he called loudly. "There is someone important here to see you."

"Who's there?" asked de Bry in French.

"The police captain," came the reply.

We looked at each other, stunned. *Police!* We didn't plan to keep the
coins, only examine and photograph them like any other artifacts from
the site. Our reasons for not advertising this find were simple: just the
sight of gold could cause a gold rush on the island, and one of our eth-
ical responsibilities was to safeguard the site from potential looting.

I leaned over a table and pretended to be looking at a map. John
unzipped the flap.

"Good afternoon," he said in French, his face stretched back in his
broadest smile.

"Good afternoon," said the police captain, looking uncomfortable.
Standing next to him was a junior officer. Both were dressed in their
formal police outfits with fezlike hats to top off the ensemble. "We

BRANDON CLIFFORD LOOKS AT THE WRECK SITE FROM THE DOOR OF THE
EXPEDITION TENT. *Barry Clifford*

would like to ask Mr. Barry if he is interested in donating money to the
police station's building-repair fund. We need to paint and plaster the
police station, and there isn't enough tax money to do it."

"We'll get back to you," said de Bry, a response that seemed to satisfy
the police captain, who saluted and left with his lieutenant.

We restrained ourselves as long as we could and then broke out in
laughter. So far I had been asked to donate to the elementary school,
an orphanage, the "educational fund" of a taxi driver, the colonial governor's house-renovation fund, and the sports club of the local army
soccer team, as well as a number of other real or perceived charities.
This time, though, I was actually elated that the people knocking on
our door were asking for a contribution. I had thought they would be
slapping handcuffs on our wrists.

"Let's get these coins back down to the wreck before something else happens," I said to de Bry.

The next day we put them into a weighted plastic bag and buried them deep in the ballast mound, exactly where they had been found, while a Discovery Channel cameraman filmed our every move just so we would have a visual record of the coins' return.

After the depressed mood of the last few days I was enjoying our lucky streak. Many who have walked beaches from Mona Island to the shores of Maine have looked for the pirate treasure of Captain Kidd. Robert Louis Stevenson had Kidd's treasure in mind when he wrote *Treasure Island*. No one, though, has ever found a single piece of gold that belonged to Kidd. But now *maybe* we had done it. Perhaps we had found pieces of gold that belonged on Kidd's ship the *Adventure Galley*, or maybe even to Kidd himself. We had touched the gold of Captain Kidd. Or so I hoped.

20

"Never a Greater Liar"

Bellomont now had to defuse a threatening scandal, one that he referred to as the "Kidd problem."

He had invested in the disastrous venture because he thought he could make a tidy profit. According to the agreement, Bellomont expected to be dividing at least £100,000. Recently though, he had heard rumors (possibly from Campbell) that Kidd was carrying as much as £500,000 in booty. So it must have come as a rude surprise to find that the man he had helped sponsor claimed to have made a total of only £40,000. *And,* worst of all, that he had likely turned pirate in the process.

Bellomont did the math. If Kidd's booty really did amount to £40,000, his take would be less than £5,000, once the expenses were covered and the money divided among the partners.

Of course, that was the amount Lord Bellomont would receive if he took the risk of interceding on Kidd's behalf. If he arrested Kidd and charged him with piracy, the numbers got much better, since as a vice admiral, he could legally claim one third of any stolen loot he received. Under those circumstances, he would make more than £13,000 and neatly rook his other partners out of their shares of the swag as well.

It looked as though treating Kidd as a pirate would solve all of Lord

Bellomont's problems. Not only would it provide him with a much needed cash infusion, it would also bolster his political profile. All parties of the government—Whigs *and* Tories—were demanding an end to piracy. They knew the importance of the East India Company to the nation's economy and were well aware of the problems caused by Kidd in the Indian Ocean. It would be beneficial for Lord Bellomont if he appeared to be tough on crime, not to mention the fact that he was under direct orders to arrest Kidd, and that he had been the lead man in the venture in the first place.

With a careful pen he wrote the very letter that Kidd received with such relief, and that would lure him into the trap. Now all he needed to do was wait.

WHEN CAMPBELL RETURNED FROM Boston to escort Kidd back to Lord Bellomont, the pirate captain was almost giddy with relief. First he handed out some of his booty to Campbell. For all of his trouble, the Boston postmaster was given one hundred pieces of eight, some pieces of muslin cloth, and a gold chain for his wife. To the man who accompanied Campbell to the *San Antonio,* Kidd gave a bale of white calico, some pieces of muslin, and a small quantity of sugar. He even presented some pieces of Arabian gold to the man who rowed Campbell and his companion between the two ships.

Kidd also offered Campbell five hundred pounds if he could convince Bellomont to obtain a pardon for him, which Campbell later said perplexed him, since he truly had no say in decisions of Lord Bellomont or the council.

Instead of going directly to Boston, Kidd backtracked to Gardiners Island off the eastern tip of Long Island. He had already deposited a large amount of cargo on Block Island, and now he anchored off Gardiners Island where he left cloth, three slaves, and a large chest of gold and silver with John Gardiner, the owner of the island. Kidd actually buried the chest and told Gardiner that it must stay there until he returned from his business in Boston.

This rendering of Kidd burying treasure on Gardiners Island is of an actual event. Although the treasure was dug up later and returned to authorities, it contributed to the belief that Kidd buried treasure in a number of locations.
Howard Pyle

En route to Boston Kidd stopped in Tarpolin Cove, near the shoulder of Cape Cod, where he unloaded more goods. On his way along the Cape, he hailed a sloop that was bound for Boston and asked the captain to transport a bag of goods containing pieces of eight, a Turkish carpet, and some clothing belonging to Mrs. Kidd. As payment, Kidd gave the surprised master a bar of gold.

After squirreling away much of his booty, Kidd took the *San Antonio* around the tip of Cape Cod and into Massachusetts Bay. It was July 2, 1699. As his sloop sailed into Boston Harbor, Kidd was nervously rehashing his alibi. The letter from Bellomont had made it clear that he had some convincing to do, but there was that one hopeful line that said, "I make no manner of doubt but to obtain the King's pardon for you and those few men you have left. . . ." That sounded good to Kidd, as did the line before it that read, "You may safely come hither, and be equipped and fitted out to go and fetch the other Ship." That "other ship" was the *Quedagh Merchant,* the one that Kidd alone knew how to find.

Kidd is universally described as "cocky" and "arrogant," so there would be no reason to think he was anything but confident as his ship docked in Boston. He'd been lulled into a false sense of security by Bellomont's cajoling letter and by the friendliness of Campbell, who had even invited the entire Kidd family to stay in his well-appointed rooming house.

The trap was about to be snapped shut.

As soon as the Kidds got settled in Campbell's rooming house, Bellomont asked for a meeting with him that lasted well into the night. Kidd tried to sell his alibi to the increasingly skeptical Irishman, but Bellomont did not accept the story of mutiny and treachery that Kidd deftly wove for him. He was no longer sure that Kidd had left money behind in the *Quedagh Merchant,* either. He told Kidd that he would have to tell his story again, this time to the council, and he ordered him to appear before them at five the next evening.

"There was never a greater liar or thief in the world than this Kidd," Lord Bellomont later declared.

THE NEXT MORNING KIDD tried to win back Lord Bellomont's favor. He sent a gold bar by courier to Mrs. Bellomont, which she refused to accept. He had already sent her an ornate enamel box containing four jewels when he corresponded with her husband from Block Island. Although she would have preferred to keep it, Lord Bellomont felt that the expensive gift might not look good to a government oversight committee and insisted that it be returned.

While he waited nervously for the council meeting, Kidd was visited by Robert Livingston. The Scottish entrepreneur asked his close friend if Kidd had struck it rich on his voyage and whether his own investment of ten thousand pounds was safe. Rumors had flown before Kidd's return, including one that said he had pulled in a total of five hundred thousand pounds in booty. Was that true?

Kidd sadly shook his head. He had not pulled in anywhere near that amount. He did reveal to his old friend a piece of new information, however. He had a forty-pound bag of gold hidden, and he would only release its whereabouts after he knew he was safe.

Livingston left Kidd and went directly to Lord Bellomont. Deeply concerned about losing his bond money, he showed up on Bellomont's doorstep in such an agitated state that the governor later recalled it in a letter to the Board of Trade:

> Mr. Livingston . . . came to me in a preemptory manner and demanded up his Bond and the articles which he sealed to me upon Kid's Expedition, and told me that Kid swore all the Oaths in the World that unless I did immediately to indemnifie Mr. Livingston by giving up his Securities he would never bring in that great Ship and Cargo, but that he would take care to satisfie Mr. Livingston himself out of that Cargo. I thought that this was such an Impertinence, in both Kid and Livingston, that it was time for me to look about me, and to secure Kid.

KIDD'S FATE TOOK A TURN for the worse at that evening's council meeting. Before the council members, Kidd retold the story of how he lost his two-year battle with mutinous crewmen, and when the council demanded that he show them his logbook, Kidd claimed that the crew had stolen it on Sainte-Marie Island. This was a common dodge among captains who were being investigated for acts of piracy, since a logbook is a record of a ship's daily activity. The fact that he had no written report to show, along with his general demeanor, made Kidd appear guilty to the council.

A council member asked Kidd to tell them what cargo he had brought back to Boston on the *San Antonio*. From memory he listed the contents of the cargo hold:

Forty Bales, containing Callicoes, Silk, Muslins striped and plain.
Five or Six Tons of refined Sugar, contained in Baggs.
About 40 lb. Weight in Dust and Bar Gold.
About 80 lb. Weight in Bar Silver

The council members must have laughed when Kidd claimed that all of this—including the forty pounds of gold—had been purchased in Madagascar by selling powder, small arms, and furniture from the *Adventure Galley*. They knew full well that there was no way such a rich haul could be gathered from the furnishings of a ship the size of a galley.

An accounting of the cargo that remained on the *Quedagh Merchant* was also given. It consisted of 150 bales of cloth, several tons of sugar and saltpeter, guns, anchors, and chunks of iron. "There is no Gold or Silver on board that he knows of," read the council proceedings.

From Kidd's manifests of both ships it seemed likely that very little booty remained on the *Quedagh Merchant*, and that it certainly did not include the thirty thousand pounds that he claimed was there.

Lord Bellomont may have breathed a sigh of relief at this point. He knew now that he did not have to risk another voyage by Kidd to retrieve the booty from the *Quedagh Merchant,* since there was no real booty to speak of. He could now get on with the job of building a case against Kidd.

It would not be difficult. The Council had given Kidd until five the next evening to prepare a written narrative of his voyage. At the appointed hour he arrived without the account but with five of his men, hoping that the council would allow them to corroborate his story. The council sternly rejected their obviously well-rehearsed testimony and demanded Kidd's written account of the voyage.

Angered by the rejection, Kidd turned arrogant. He refused to tell the council where the *Quedagh Merchant* was in Hispaniola, and then even denied that the ship he had left there was truly the Indian Ocean trader.

The council dismissed Kidd and allowed him twenty-four hours to produce a written narrative. When the appointed hour arrived, Kidd insisted that he needed more time. Another twenty-four hours was given, and this time Kidd did not show up at all.

Concerned that he was trying to escape from Boston, the council ordered Kidd's arrest. Later that day, on July 6, Kidd was spotted outside Bellomont's house, wearing a sword. As the constable approached he attempted to draw it. A scuffle ensued as Kidd wrestled to open Bellomont's door. In a moment he stood face-to-face with the lord himself.

Bellomont had no mercy. He ordered that Kidd be immediately stripped of weapons and taken to the Boston jail. Then he insisted that Kidd's lodging be searched and all of the treasures remaining be seized. He also ordered guards to be posted on the *San Antonio* to keep the ship from being looted by the citizenry of Boston. Then the remaining members of Kidd's crew were rounded up and jailed with their captain.

Lord Bellomont must have felt both lucky and self-righteous on this day. Not only had he just arrested England's most wanted man, he had secured a substantial fortune as well.

21

THE FIERY DRAGON

THE STORM THAT UNCOVERED THE gold coins on the wreck site had revealed far more than gold. It had exposed evidence that would change the direction of our search.

Just before John de Bry found the coins, he was fanning silt away from the beams of the hull so he could examine the ship's construction. The storm had stirred up the bottom of the bay, clearing mud from the wreck in such a way that parts of the hull could be easily inspected.

The appearance of the wood grain told de Bry that the hull of the ship we were looking at was made of oak, which matched the record of the *Adventure Galley*'s. As he cleared away more of the silt, other aspects of the ship's architecture came into view and he could see where the floor timbers joined the futtocks, the curved timbers that form the ribs of a sailing ship's hull. What de Bry saw here came as a shock: "We're not diving on the *Adventure Galley*."

"What?"

"The ballast mound belongs to another ship."

I was deflated.

We retired to his bungalow, where he patiently laid out his case. The way the futtocks were fastened to the floor timbers suggested a Dutch construction rather than English, he said. While English shipwrights

BARRY CLIFFORD AND JOHN DE BRY PLAN A FUTURE DIVE USING THE MOSAIC MAP OF THE WRECK SITE, THE ONE THEY WOULD DISCOVER WAS THE *FIERY DRAGON*. *Paul Perry*

used wooden dowels to join floor timbers and futtocks together because it allowed for quick dismantling and replacement of damaged or rotten timber, the Dutch generally preferred either iron fasteners or a combination of iron fasteners and dowels. Further, certain characteristics of the porcelain told him that the shards that covered the ballast mound were of the style made after 1700, possibly around 1720 or so, more than two decades after the *Adventure Galley* had sunk.

I was disheartened. Although the site was almost certainly that of another pirate ship, it didn't help me much in my search for Captain Kidd. The consolation was that we were close, but by that time we had no other choice but to break camp and prepare for the long voyage home. De Bry and I agreed that he should stop in France so he could continue the process of dating and identifying the wreck. He was going to focus on the porcelain we had found, particularly the double-headed eagle that decorated the surface of some of the shards. De Bry also contacted Ken Kinkor back in Cape Cod with the new facts concerning the shipwreck's identity.

"Do you know which European-built ships besides the *Adventure Galley* have sunk in Sainte-Marie harbor?" de Bry asked. "They would have sunk between 1700 and 1725."

The pirate historian's response was quick. "The only one recorded was the *Fiery Dragon* in 1721," said Kinkor. "She was commanded by William Condon."

Kinkor said he would put together a history of the *Fiery Dragon* and Captain Condon for us so we could understand what we were dealing with.

Because so much had now changed, I felt certain we would be returning for a third expedition to Sainte-Marie. We secured storage space for much of our equipment from Fifou Mayer so we would not have to transport it back to the States and risk damage or theft by the airlines, and I hired Gregory to watch the wreck site while we were gone.

"How long will you be gone this time?" he asked.

Truth was, I didn't know. Our search for the *Adventure Galley* had suddenly gone down a different road, one that might be seen as a dead end by the producers at the Discovery Channel. If that were the case, the filming of the documentary might be canceled and a third expedition would be in the very distant future if at all.

"I hope it's just a few months away," I told Gregory.

He nodded sincerely. "I will watch things for you while you are gone," he said.

BACK IN PROVINCETOWN, KINKOR had pulled together information on the *Fiery Dragon* and its colorful captain, William Condon, and written a brief history about him and the ship that became one of the most successful pirate vessels in history. Kinkor called him a "pirate's pirate," because he was elected captain by his crew, instead of

GREGORY, THE LOCAL SECURITY GUARD, DILIGENTLY WATCHES THE
EXPEDITION TENT, OFTEN STAYING THERE TWENTY-FOUR HOURS A DAY.
Barry Clifford

appointed by a group of investors like Kidd. Compared with Kidd, this
made Condon relatively "democratic and egalitarian," according to
Kinkor. Condon was also fearless in combat and was not afraid to take
chances, unlike Kidd. Also, while Kidd was ill-tempered and blustering,
Condon was able to negotiate well among differing factions of his crew
as well as outsiders.

When I read the memo I began to relax about our situation.

Known as Condent, Congdon, Connor, and Condell, and by first
names Christopher, Edmond, John, and William, the brigand has
been nicknamed "Billy One-Hand" by modern "piratologists." He
and his men were an unusually hard-bitten pack of seawolves who
ran up a long string of daring robberies.

Anticipating the arrival of British warships at the Bahamas in February 1718, Condon and ninety-eight other rough "old Standers"—all veterans of other pirate crews—banded together and left Nassau in a heavily armed sloop named the *Dragon.* Shortly afterward, an irate crew member barricaded himself in the hold, threatening to ignite the powder magazine in revenge for a beating he had received earlier. Some advised prying up deck planks and tossing down grenades to flush him out, but Condon leaped into the hold with a pistol in one hand and a cutlass in the other. His arm was shattered by a bullet from the darkness, but he returned fire and the mutineer died.

After enduring amputation of his arm, Condon was promoted from quartermaster to captain for his bravery. After a short spell of robbery that took them through the Lesser Antilles in the Caribbean, the crew voted for the shores of West Africa. Setting sail for the Cape Verde Islands in mid-March, Condon took the *Alexander,* a British merchant ship laden with wine from Madeira, which he and his crew then plundered and mounted with twenty-two guns. Some forty disgruntled merchant seamen from various other prizes had already joined the gang.

Billy had plenty of sympathy for such men. After taking about twenty more vessels in one fell swoop, Condon "took upon him the Administration of Justice, enquiring into the Manner of the Commanders' Behaviour to their Men, and those, against whom Complaint was made, he whipp'd and pickled" (doused with seawater).

At Saint Jago in the Cape Verdes, a large Dutch ship was spotted. Though seriously overmatched, Billy One-Hand promptly came alongside, poured in a devastating broadside, "laid her aboard," and captured her without further ado. Naming this ship the *Flying Dragon,* Condon took her over for his own, releasing both the *Alexander* and his old sloop the *Dragon.* He then ranged both sides of the South Atlantic, sweeping the seas clear of English, French, Portuguese, and Dutch shipping.

Not all victims were looted. One such was John Spelt, commander of a slave trader called the *Wright Galley*. Spelt, like Condon, was from Plymouth (as was Sam Bellamy, pirate captain of the *Whydah*), and although Condon kept the *Wright Galley* in company for weeks, he treated his fellow townsman "very civilly," presenting him with sundry stolen goods as amends for his trouble.

Not so lucky was a Dutch East-Indiaman of twenty-six guns. Condon's gunnery was up to its usual standard; the Dutch captain was killed in the first broadside and the ship easily taken.

Off Brazil, the pirates happened upon a Portuguese man-of-war of seventy guns. When the Portuguese hailed the rovers, the answer came back that they were a slave ship out of London bound for Buenos Aires. Fooled, the Portuguese manned their shrouds for a complimentary cheer. As the *Flying Dragon* passed, she suddenly erupted with cannon and small-arms fire. After a ninety-minute duel, Condon casually sailed off unscathed, having killed more than forty men in the warship.

Continuing southward, he took *La Dauphine,* a French ship of eighteen guns heavy with wine and brandy, which he carried with him into the River Plate for a leisurely plundering toward the end of 1718. While hunting wild cattle ashore, some of his men were captured by the crew of a Spanish man-of-war, but they talked fast and managed to get themselves released.

Hearing that some shipwrecked pirates had been lately executed in Brazil, Condon's crew reportedly retaliated by thereafter treating all Portuguese captives "very barbarously, cutting off their Ears and Noses." Condon then recrossed the Atlantic and took the *Indian Queen,* Captain Hill, off the coast of Guinea.

At Luengo Bay on the southwest coast of Africa, a Dutchman of forty-four guns and the English ship *Fame* were sighted at anchor. As the pirates approached, both ships cut their cables and ran themselves ashore. The *Fame* was wrecked, but the marauders managed to refloat the Dutchman. The pirates took this vessel over and named her once again the *Fiery Dragon*. Like her namesake, she had teeth. She was fitted with forty cannon, twenty brass swivel guns, and even three-man portable Cohorn mortars, which could lob exploding shells onto the decks of opposing ships. With a crew of about 320 men and small arms for twice that number, she was as well manned as she was well armed.

In late October or early November 1719, the *Georges* was taken off the west coast of South Africa, and a month or two later Condon looted the *Prince Eugene* as she sailed below the Cape of Good Hope.

The next victim came off the Cape of Good Hope in February 1720 in the form of the twenty-four-gun, four-hundred-ton *Maison d'Autriche ("House of Austria")*, homeward bound to Oostende (in modern-day Belgium) from Canton, China. Her captain, a former English East India commander named James Nash, readily turned over a hidden parcel of gold and other valuable cargo to his captors and was released relatively unscathed.

After taking yet another Dutch East-Indiaman, the marauders made for Madagascar. In those days, Île Sainte-Marie held a high reputation as a haven for villainy and a place from which to stage attacks against the incredibly rich pilgrim ships sailing between India and Jeddah, the port of the holy Arabian city of Mecca. At this "pirates' paradise" Condon recruited among the local population of hardened career pirates as well as local blacks. Most pirates welcomed black recruits and depended on the friendship of these islanders for both resupply and sanctuary. Condon left port on the southwest monsoon bound for the rich pilgrim ships traveling between India and Arabia.

Near Bombay, on August 18, 1720, the raiders found exactly what they were looking for—a five-hundred-ton Turkish-owned pilgrim ship homeward bound for Surat from Jeddah. The passengers and crew wisely offered no resistance and were safely put ashore near Surat. This was a fabulously wealthy prize laden with gold coins, drugs, spices (such as frankincense, saffron, and myrrh), calicos, and silks "worth Twelve Lakh of Rupees"—perhaps 150,000 English pounds sterling (estimated at roughly $375 million today). Given that just *two* pounds sterling was an ordinary sailor's pay for an entire month, Condon's crew must have been contemplating a pleasant retirement as they returned to their base on Île Sainte-Marie.

During the seventeenth and eighteenth centuries, piracy flourished with aid and comfort from dishonest merchants and politicians. Pirates often bought firearms, gunpowder, and liquor at ridiculously high prices—for which they paid with stolen goods at ridiculously low valuations, or even in good hard cash. And so many a bloodstained piece of eight was rubbed clean in the silk pocket of a corrupt merchant.

Thus it was that the *Coker Snow* of London arrived at Sainte-Marie in October 1720. Although her captain was one Richard Taylor, she was under temporary command of a Captain Henry Baker.

Baker and Taylor were ashore trading when, on October 13, they spotted two large vessels working their way into the port. They hoisted sail but were unable to clear the harbor in time. Their visitors proved to be the *Fiery Dragon* and her prize.

After getting an account of the *Coker*'s cargo, Condon took out five hundred gallons of wine and other liquor, which he was careful to pay for. Although the desperados had made their fortune, they needed legal sanctuary in order to retire comfortably. The best way to obtain such in those days was to bribe a corrupt governor to issue a pardon, and the *Coker* represented a conduit to just such a governor. The pirates had heard that Governor Joseph Beauvollier de Courchant of the French island Île Bourbon (modern Réunion) was offering pardons to any pirate giving up his trade. While many of the *Fiery Dragon* pirates opted to settle on Sainte-Marie or elsewhere on Madagascar, Condon and others voted on October 16 to send Baker and five men to Réunion Island to investigate the matter—while keeping Captain Taylor and five of his merchant crewmen as hostages.

The *Coker* dropped anchor off Réunion on November 12, 1720, and Captain Baker transmitted the pirates' request for a pardon. The governor was more than happy to oblige; he was under orders from Versailles to make every attempt to persuade pirates to surrender and give up their nefarious trade. De Courchant convened the provincial council and, with their advice, drew up a proposal for Condon and the pirate company. They were to surrender themselves, their weapons, and their ammunition to the authorities at Réunion within four months, and to surrender or destroy their warships. In return, each pirate would receive a pardon, and, with the payment of a small fee, the right to settle on the island. Each of the pirates could also bring in one "Negro slave"—provided that these were not "warriors" who had fought side by side with the pirates. Eventually this was to cause trouble, as there were 60 blacks among the crew in addition to the 135 pirates who had requested the pardon.

The only opponents of the plan were missionaries of the Lazarist order who quite reasonably pointed out that the pirates were being allowed to enjoy the fruits of their sin scot-free. Versailles, however, took up a much more pragmatic attitude; the Lazarists were characterized as "troublemakers," and de Courchant was ordered to exclude them from the island's governing council.

The *Coker* returned to Île Sainte-Marie on November 26 with the "Act of Grace" from Réunion, but the pirates were taking no chances and demanded that she go back for a special "Act of Indemnity." They all remembered the fate of those among Kidd's crew who had surrendered under a general pardon—only to be hanged later under a technicality—and so they wanted to make sure their agreement was ironclad. The *Coker* accordingly set forth again to Réunion and obtained the desired act, with which she finally returned on December 27. When the matter was then put to the men for a vote, forty-three agreed to accept the pardon, while forty men decided to take their chances on Île Sainte-Marie; the remainder had already shipped out or had fled by boat across the narrow strait to the main island of Madagascar.

According to the agreement, the pirate ships were to be surrendered to the French authorities. Many of the pirates were disabled by sickness, however, and there weren't enough hands to man both vessels. To prevent the "irreconcilables" from jeopardizing the indemnity deal by a return to piracy, both ships were destroyed. For years thereafter, drugs, spices, and chinaware were seen lying in heaps on the beach, exposed to wind and weather.

For their passage each of the thirty-two white pirates deciding to surrender promised to pay "50 pounds sterling and a slave," and on January 9, 1721, the *Coker* sailed with its cargo of "retirees" for Réunion.

Upon the safe arrival of the vessel at Réunion, there was, as could have been expected, trouble regarding payment to the *Coker*. The pirates appealed to the governor regarding the terms of the passage. The governor settled the dispute by compelling the pirates to pay the passage money to Baker but not the "slaves"—who convinced the governor that they were indeed "warriors" and full-fledged members of the crew.

The pirates were warmly welcomed by the residents of Réunion and, in the style of Errol Flynn and other big-screen buccaneers, Billy One-Hand lived "happily ever after." Shortly after his arrival, he was even asked to negotiate the release of the Portuguese viceroy of Goa, who had been taken by the pirates Taylor and La Bouche ("the Buzzard") in April 1722. Condon was successful in mediating a ransom, and the pirates left without looting the island. The islanders were extraordinarily grateful and did everything they could to make the pirates welcome. Billy One-Hand is even said to have married the governor's sister-in-law. Nevertheless, some twenty of the pirates, apparently "homesick," went back to Europe in

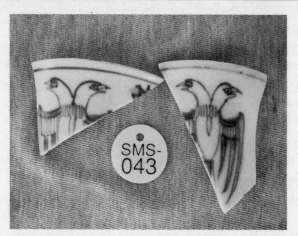

EXAMPLES OF BROKEN PORCELAIN IN THE DOUBLE-EAGLE MOTIF, SPECIALLY DESIGNED FOR THE HAPSBURG FAMILY THAT RULED WHAT IS NOW BELGIUM. THIS DESIGN HELPED PROVE THAT THE CLIFFORD EXPEDITION HAD FOUND THE *FIERY DRAGON,* ONE OF THE MOST SUCCESSFUL PIRATE SHIPS IN HISTORY. *John de Bry*

November 1722, and Condon followed them a year later. He settled in Saint-Malo, on the Normandy coast of France, where he became a thriving merchant and shipowner known locally for his "honor and probity."

KINKOR'S THOROUGH SCHOLARSHIP SHOWED the importance of our accidental find. As he pointed out at the end of his memo:

While the discovery and identification of the *Adventure Galley* wreck site would clearly be the most historically-significant accomplishment of the team, *The Fiery Dragon* wreck site will ultimately prove to be more fruitful from a strictly archaeological standpoint. As can be seen in the summary above, this pirate crew's robberies were far more numerous and extensive than were those of Kidd's crew. In fact, the treasure-trove of artifacts is completely unprecedented insofar as the potential range of cross-cultural materials is concerned. Given the historical circumstances of the abandonment of the two vessels, it can also be confidently expected that the quantity of artifacts on *The Fiery Dragon* site will be much larger than on the *Adventure Galley* site.

IN FRANCE DE BRY made a discovery that finally convinced us we'd found the *Fiery Dragon*. Curious about the double-headed eagle on many of the porcelain shards, he showed pictures of the fragments to Jean-Paul Desroches, the head curator of the Musée des Arts Asiatiques-Guimet (Guimet Museum of Asiatic Art) in Paris. A world-renowned authority on Chinese export porcelain, Desroches was shocked to see the double-eagle design on porcelain found in Madagascar.

"What the hell is that doing there?" asked Desroches, puzzling over the design on the photograph.

"It's from a sunken pirate ship," said de Bry.

After examining a list of ships that the *Fiery Dragon* had robbed, de Bry and Desroches concluded that the *Maison d'Autriche* was the most likely source of the china. Bound for Belgium from Canton, the ship was loaded with a cargo of porcelain for the Hapsburg family. As de Bry had suspected and Desroches confirmed, the double-eagle design was specially made for the Austrian royal family, which then ruled what is now Belgium. According to Desroches, only two or three unbroken pieces of this china are in existence today.

I WAS BEGINNING TO like the possibility that we'd found the *Fiery Dragon,* but de Bry cautioned me. "Barry, you should keep an open mind," he said. "For all intents and purposes, the two ships sank in the same spot. We don't really know which ship we're working on. It might be both for all we know."

Were we exploring the *Adventure Galley* or the *Fiery Dragon?* The fact was, we just didn't know.

22

As Good as Hanged

 Behind the stone walls of Boston's prison, Kidd put pen to paper and wrote the narrative of his voyage. He got right to the heart of his alibi from the first sentence in a narrative largely written in the third person: "That the Journal of the said Captain Kidd being violently taken from him in the Port of St. Marie's in Madagascar, and his life many times being threatened to be taken away from him by 97 of his men that deserted him there, he cannot give that exact Account he otherwise could have done, but as far as his memory will serve is as followeth."

He went on to tell his side of the story, much the same he had told to the council a few nights before. He conveniently omitted the violent murder of William Moore from his narrative and instead related the tale of a good captain forced to go bad by evil influences beyond his control. To believe Kidd's narrative would be to believe that he was a victim of the greed and avarice of others. He admitted to taking both the *November* and the *Quedagh Merchant* but said they were fair game because they had shown French passes when he asked about their nationality.

He claimed that he wanted to arrest the pirate Robert Culliford when he arrived on Sainte-Marie, but his crew refused the order to do so, saying they would rather kill him than arrest a fellow pirate.

Kidd declared that he was just like his partners, the East India Company and even England—a victim, not a perpetrator, of piracy.

KIDD HOPED GOOD THINGS would come of his narrative, but they didn't. On July 17 the council resolved that their prisoner should be moved to Stone Prison, a more secure facility. There he was put in irons "and company kept from him." Clearly, Bellomont did not want this high-profile prisoner to escape, and neither did he want him talking to anyone about those involved in funding his venture.

In the meantime, Lord Bellomont penned a quick letter to the Board of Trade in London giving them the news that he had captured Kidd and several of his men.

> It will not be unwelcome News to your Lordships to tell you that I secured Captain Kidd last Thursday in the Gaol of this Town with five or six of his men. He had been hovering on the Coast towards New-York for more than a fortnight, and sent to one Mr. Emot to come from New-York to him at a place called Oyster-Bay in Nassau Island not far from New-York.

Bellomont revealed in his letter that he did not like Emott, in part because he was a Jacobite (Jacobites gave their allegiance to James II, the monarch ousted by William and his followers). His friendship with Benjamin Fletcher, the crooked governor of New York, was also a factor. Emott may still come in handy, he said, because this was an opportunity to bring Kidd in peacefully and quickly: "I writ a Letter to Captain Kid inviteing him to come in, and that I would procure a pardon for him, provided he were as innocent as Mr. Emot said he was," Bellomont wrote. "Kid landed here this day Seven night; and I would not so much as speak to him but before Witnesses. I thought he looked very guilty. . . ."

Bellomont wanted to know from the Board of Trade what he should do with Kidd. He was concerned about keeping such a high-profile criminal in the Boston jail system where another pirate had recently escaped "with the Consent of the Gaoler as there is great reason to believe." Lord Bellomont did not want to lose his prize prisoner or his chance to prosecute him.

But there was another problem with keeping Kidd in the colonies. Many English officials wanted Kidd hanged—if for no other reason than fearing for their own heads should Kidd tell the wrong story. In the colonies, however, the death sentence could not be carried out.

"As the Law stand in this Country a Pyrate cannot be punished with Death," wrote Lord Bellomont. "Therefore I desire to receive orders what to do with . . . Kidd, and those Men of his I have taken."

KING WILLIAM WAS DELIGHTED to hear that Kidd had been apprehended. The Admiralty was ordered to send a ship immediately to Boston. The *Rochester* left port on September 27, 1699, but returned on November 6, beaten and battered by a savage winter storm. Wasting no time, the Admiralty sent another ship, the *Advice,* under the command of Captain Robert Wynne. Facing rough seas and uncertain weather, the *Advice* pressed on to Boston to retrieve Kidd and what remained of his crew.

Bellomont, meanwhile, was busy building the case against Kidd. He ordered statements taken from Kidd as well as everyone he had come into contact with since reaching the American coast.

From the deposition of John Gardiner, owner of Gardiners Island, he learned that two sloops had anchored next to Kidd's sloop and taken several bales of goods and other things off of the *San Antonio.*

From the deposition of a Captain Nicholas Evertze he learned the bad news that the *Quedagh Merchant* had been burned and sunk. Bellomont had still been thinking of mounting an expedition to find and retrieve the ship, hoping to claim as much as seventy thousand pounds from the remaining cargo. Now he had to abandon that planned search.

From Joseph Palmer, one of Kidd's crewmen, he learned the most damning information of all. Palmer told of Kidd's murder of William Moore. Bellomont must have been delighted to discover that his pirate prisoner was also a murderer, and he even requested a pardon for Palmer based on his cooperation with authorities. There was always a chance that Kidd could squirm out of a conviction for piracy, especially with the French passes that proved part of his story. It would be harder to convince a jury that this murder was justified.

Kidd, as far as Lord Bellomont was concerned, was as good as hanged.

KIDD WOULD SPEND TWENTY-TWO months incarcerated before his trial in England on May 8, 1701. The first eight months of that interminable period would be spent in jails in Boston, where few people were allowed to speak to him as he languished in his cell. That included even his wife, Sarah. On July 25, 1699, she wrote to Bellomont requesting an opportunity to visit her husband in jail:

> To his Excell'cy the Earle of Bellomont,
> The Peticion of Sarah Kidd humbly Sheweth
> That Your Petitioners husband Capt. Wm. Kidd, being committed unto the common Goale in Boston for Pyracie, and under Streight durance, as Alsoe in want of necessary Assistance, as well as from Your Petitioners Affection to her husband humbly pray's that your Excell'cy and Councill will be pleased to permit the sd. Sarah Kidd to have Communications with her husband, for his reliefe; in such due Season and maner, as by our Excelle'y and Councill may be tho't fitt and prescribed, to which Your Petitioner shall thankfully conforme herSelfe and ever pray etca
> Sarah SK Kidd

Sarah's petition was denied. Her husband remained in solitary confinement.

AT THE HEIGHT OF a freezing winter, the *Advice* made its way slowly across Boston Harbor. It was February 2, 1700. Captain Wynne and his sailors had battled storm after storm in crossing the icy sea. Now they tied up their ship on the docks of Boston, anxious because they knew they had to turn around and head back to England. Captain Wynne paid his respects to Lord Bellomont and then prepared his ship for the treacherous return journey.

On March 1 Kidd and thirty-two other prisoners were led up the gangplank of the *Advice*. Kidd was chained and shackled as he walked onto the ship's wooden deck that freezing cold day. Straightaway he was led to a tiny cabin in steerage from which he did not emerge during the remainder of the voyage.

It was a grim trip for Kidd. The waters of the Atlantic were rough, and his cabin was a freezing pit, surrounded by the ocean's cold water.

Close to England, things only got worse. Try as he might, Captain Wynne was unable to bring the *Advice* into the English Channel. The weather was so fierce and the waves so high that he put in at Lundy Isle instead to wait out the heavy weather. A letter was sent telling the Admiralty where they were.

A royal yacht was dispatched to tell the captain and crew of the *Advice* that Kidd was to speak to no one and send no messages. Already he had been prevented from talking to the crew or any of the other prisoners on board. He received his meals by the hands of the captain's cabin boy, who did not speak to him. Now Kidd knew that he was in for another lengthy stint of loneliness.

Kidd's solitary confinement was not an intentional punishment. Rather it was a form of political cover-up. Four members of the Whig party had provided financial support to Kidd's venture. If Kidd were to talk about the extent to which these government officials had been involved, it could prove to be an embarrassment at a time when it would hurt the most. The country's war with France had just ended, and a bitter fight raged between Whigs and Tories over continued support of a large standing military. The Whigs wanted to substantially reduce the size of the military, while the Tories backed a considerable increase.

The Kidd scandal gave the Tories a chance to make the Whigs look foolish, if not criminal. After all, it was Whig politicians who had supported Kidd's private venture as a "pirate hunter." Now the Tories could point out that the privatization of the military hadn't worked well at all. And of course they would be right.

THE *ADVICE* FINALLY MADE it to Greenwich. After several months in confinement and more than two months in the icy cabin, Kidd was physically and mentally fragile. So much so that John Cheeke, the Admiralty marshal, was alarmed at Kidd's appearance and the cloud of depression ranging over him. He told Cheeke that he hoped he would die by firing squad instead of the hangman's noose. When he asked his jailer if he could be slipped a knife so he could slit his wrists, he was refused.

A barge was sent to pick up Kidd on April 14 and bring him to London. Too weak to walk unaided, the prisoner was transferred to a sedan chair and transported to the Admiralty building. He was delivered to the boardroom, where he faced four staunch members of the Tory party. What they were hoping for was information that would be damning to the four members of the Whig party who had helped fund his venture. Had Kidd been willing to present it, they may well have granted him a pardon. But something wasn't quite right with Kidd. One account of the meeting says that he was drunk, but historians have speculated that he may have suffered a minor stroke. Whatever the case, Kidd delivered a lengthy ramble that made little sense to the politicians. When the hearing was over, the Tories were somewhat disappointed in their star witness. One of the members remarked, "I had thought him only a knave. I now know him to be a fool as well."

When Kidd was questioned again the next day, he made no better a showing.

The members of Parliament were finished with Kidd. He had proved to be worthless to their cause and would now be treated like a common criminal. Worse even, Kidd was transferred from the Admiralty jail, the Marshalsea, to Newgate Prison. Newgate was a horror, a place so awful that people on the streets would often refuse to even look at it.

It was decided that Kidd, classified a traitor by judges of the Whig party, could legally be held incommunicado. He became seriously ill with fevers and chills, so much so that the Admiralty relented and allowed him to have bedding and new clothing. Kidd asked to be allowed to talk to other members of the prison population. His request was denied. Instead they let him have occasional visits from his uncle, a man named Blackborne, his wife's aunt, Mrs. Hawkins, and an old friend, Mathew Hawkins. Once a week or so he was allowed a visit to the prison chapel. Other than that, Kidd was alone.

And so he lived until his trial more than a year later.

PART III

THE THIRD EXPEDITION

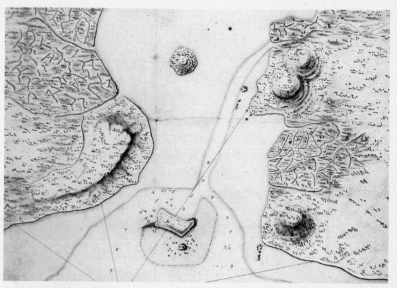

Reader, near his tomb don't stand,
Without some essence in thy hand;
For here Kidd's stinking corps does lie,
The scent of which may thee infect:
He base did live, and base did die.

— "ELEGY ON
THE DEATH OF CAPTAIN WILLIAM KIDD"

23

FRIDAY THE THIRTEENTH

THE *ADVENTURE GALLEY* OR THE *FIERY Dragon?* That was the question.

The evidence was strong for both of them. Maps and pirate depositions told us that the *Adventure Galley* had sunk in that very spot. Historical records told us that the *Fiery Dragon* had sunk in that spot, too. Both ships had been burned. To add to the confusion, a further examination of our side-scan sonar results had revealed the outline of two more sunken ships about fifty feet west of our wreck site. They appeared to be identical to the ones in our primary site: a large ship with a small ship behind it, as though it had been towed in.

All four ships were victims of nearly identical circumstances in identical spots.

When I told Michael Quattrone at the Discovery Channel that we may be on to the *Fiery Dragon* as well as the *Adventure Galley,* he became excited. One of these big ships was the *Adventure Galley,* while the other was the *Fiery Dragon,* one of the most successful pirate ships in history. All we had to do, he said, was sort them out.

He quickly approved a third expedition, slated for October 2000.

IN SEPTEMBER WE BEGAN to contact the Malagasy ambassador in Washington, D.C., again to get the necessary visas and permits. This time we needed both filming and excavation permits, not just the filming permits we'd had on the second expedition.

We didn't expect any problems. Surely after two expeditions we had demonstrated our good intentions to Malagasy officials. Even with the hassles we'd had in June, our relationship with Madagascar's baffling bureaucrats had improved. This time it would be a piece of cake.

But when my office contacted Ambassador Zina Andrianarivelo-Razafy, they received a frosty reception. The ambassador told us that there would be no further expeditions to search for Captain Kidd's ship or any other ship. Period.

At first I thought there was some kind of mistake, that they had confused my expedition with another. When a second call resulted in the same response, I became angry. We had abided by all regulations and requirements. What could possibly have happened in the interim to change their minds?

I immediately went into high gear, contacting everyone I thought could plead our case. But even letters from our congressmen were met with ice from the ambassador.

I was baffled. What happened?

John de Bry was puzzled, too, but as always he had a good Plan B.

"I met President Didier Ratsiraka's daughter at a meeting of UNESCO last year," he said. "Nice lady. We got along great. I'll give her a call and see if she can give us any help."

"I will gladly help you," Annick Ratsiraka told de Bry when he called. "I am a good friend with the Malagasy ambassador. In fact, I am responsible for him getting his job. But ever since he took the job in Washington he has become, how do you say? Too big for himself."

The seesaw battle between the president's daughter in France and the president's ambassador in Washington continued for more than two weeks until the reason for keeping us from going finally emerged: someone in the archaeological community wanted me kept off the island.

"There is somebody in your own business that does not like you," said Annick. "I will try to work around it, though. The ambassador owes this to me."

Eventually the tug-of-wills escalated to a tug-of-war. On September 28 the trip was canceled. Concerned about its reputation in the diplomatic arena, the Discovery Channel decided to pull the plug.

I was livid. We were on the brink of a great discovery, and a bureaucrat was stopping the expedition and would not even give us the courtesy of telling us why. It made no sense.

I called each member of the crew with the bad news.

As soon as I put down the phone, I received a call from the Discovery Channel producer. Quattrone had insisted that the trip be scheduled. As low as I had just felt, I suddenly zoomed to an emotional high.

Six days later the trip was canceled again. De Bry placed another call to Annick, who ordered the Washington ambassador to issue permits. "I don't know what's wrong with him," she said. "I have seen him arrogant before, but never *this* arrogant."

Once again Ken Kinkor gathered the passports of the expedition team and sent them to the Washington embassy, where each would be stamped with a "special permit." Then he rebooked the airplane reservations and everyone repacked their bags.

"We have our permits," I told Quattrone. "We're ready to go."

The day before we left, our passports came back with a very special permit indeed. Underneath a stamp from the embassy in Washington and the signature of Andrianarivelo-Razafy was a paragraph in French that allowed us to film the wreck site but not physically disturb it.

I wanted to throw the passports with the "special permits" into Provincetown Harbor.

I contacted de Bry immediately with the distressing news. He called back a few moments later with news that provided only scant relief. "Annick says not to worry. She said we will have the permits to excavate when we get there. She'll have the minister of culture issue them, and we'll have a signing party. She said to bring a camera so it can be part of the film."

I hung up feeling only slightly better. The thought of flying a dozen divers and archaeologists halfway around the world without the permits to get their job done was not comforting. On the other hand, the president's daughter had said that the permits would be ready when we arrived. She even promised a party. *I can use a party right now,* I said to myself. The last month had been hell.

Today was Wednesday, October 11. Tomorrow we would fly out of Boston for Paris and then immediately on to Antananarivo, the capital of Madagascar. For just a moment I was overcome by superstitious sentiment that roiled up from some forgotten spot in my psyche.

We would be landing in Madagascar on Friday the thirteenth.

WITH A LOUD *THUMP,* the customs official branded another passport with the rubber-stamp imprint of the country of Madagascar. Then he reached his hand through the tiny window in his glass booth and asked for the passport of the next person in line.

"So, Dr. de Bry, I see that this is your third visit this year to Madagascar," said the smartly dressed customs officer. "Why have you returned to our happy island?"

"I have come to look for the *Adventure Galley,* a sunken pirate ship on Île Sainte-Marie," said de Bry. "It's all right there on the passport."

"I see," said the customs officer, reading the special visa. "How many are there in your group?"

"There are a dozen," said de Bry after a moment of calculation. "They're back there somewhere."

The customs officer rose for a moment, peeking over the wall of his booth at the 311 passengers who had just crossed the tarmac from the Paris-to-Antananarivo flight.

"I wish you a successful trip," he said, raising the rubber stamp high above the well-worn passport and bringing it down hard. *Thump.* "Next," he shouted.

And so it was for the next eleven members of the expedition. A quick question, a sharp *thump,* and we were on our way through customs and into the cool night air of Madagascar's capital city.

I tried to ignore the date. Instead I focused on Annick Ratsiraka's promise that we would receive our permits to excavate immediately. "This will be the best expedition yet," I said to John.

"No doubt," he agreed.

Dragging our piles of luggage, we climbed into several taxis that would take us to the Tana Plaza Hotel, where we planned to decompress for two days before leaving for Sainte-Marie. The shabby Renault taxi cabs were spewing fumes of leaded gasoline, their cockeyed headlights providing shaky images of street life in the capital city of the fourth most poverty-stricken country in the world.

"I can't wait to get those permits and start excavation," said de Bry impatiently after we arrived at the hotel. "Tomorrow let's get to Annick's office by noon and get everything in order."

WE FOUND ANNICK'S OFFICE in a tall white apartment house across the street from the American embassy. Satellite dishes figured prominently in the skyline, and the street was lined with new cars from all over the world. Armed guards in fatigues manned a barricade, an iron bar across the road that they raised when cars approached.

De Bry, Paul Perry, and I climbed the circular stairs to Annick's office apartment and knocked on her door. Three assistants, thin men with broad smiles, ushered us into a stark room with white walls, a black checkerboard floor, and big picture windows that looked onto the city.

"Here you are, finally," said Annick, entering the room from her office. Although it was the middle of the day, she was wearing a tight blue cocktail dress and high heels. Motioning us to follow, she walked back to her office, chattering about the beauty of the day and how lucky we were to be there then.

"The jacaranda trees are blooming," she said, referring to the mas-

ANNICK RATSIRAKA, DAUGHTER OF MADAGASCAR PRESIDENT DIDIER
RATSIRAKA, AFTER A PLEASANT BUT NOT VERY HELPFUL FIRST MEETING.
Paul Perry

sive trees that were blossoming with lavender-colored flowers. "People
live for this time of year."

The small talk continued. There were no pictures on the walls of
her office, and only a few knickknacks, from Egypt. There wasn't even
a file cabinet or permanent telephone. Annick Ratsiraka seemed to
carry everything she needed with her in a tidy briefcase, a cell phone,
and a compact Sony computer.

"I spend a lot of time in France," she said. "Even though my father
is president, there are some things here that I am frightened of. I am
terrified of getting sick in my own country because medical facilities
are so few and so bad here. I also find it terrifying that women can die
of birthing in my country just because they live far from the city."

We told her of some of the suffering we had seen on the island, especially a little girl who had been burned badly in a family kerosene fire and now lived in painful misery because of inadequate medical care. Annick nodded. "The problems here are so big that we don't know where to start," she said, a lovely smile gracing her face.

Sensing an opportunity to change the subject, I asked her about the excavation permits.

"I don't know what is happening with those permits," she said angrily. "I see this project as an opportunity for our country, and I requested that they be done, but . . ."

What was clear was that there would be no signing ceremony today as she had promised. And probably none tomorrow. Things were on hold, she declared. The permit to excavate the *Adventure Galley* site needed the signature of three cabinet ministers, she explained. As of this moment it had been signed by none of them.

I looked at Perry and de Bry as the news came to us from across the smooth ebony desk. Their faces began to melt like wax candles in an oven. I'm sure mine was doing the same. The expedition had run aground.

I remember little of our conversation after that. I know that we all struggled to keep the small talk upbeat, and that we chatted about the medical aspects of the expedition, including the several boxes of hospital supplies we were bringing to the medically deprived island of Sainte-Marie. De Bry spoke about the importance of excavating pirate artifacts that would enrich the cultural knowledge of Madagascar, and about our hopes to build a museum on the tiny island to house the artifacts that would come right out of its own bay. But to do this, he said, we needed an excavation permit and fast. A few blocks away, nine members of the expedition sat in the Tana Plaza Hotel waiting to work, and the meter was running.

"I will call the minister of the interior right now and see what is happening," said Annick, opening her cell phone and scrolling through

JOHN DE BRY WITH THE VICTORY CIGAR THAT HE DIDN'T LIGHT FOR SEVERAL WEEKS. *Paul Perry*

the phone book. She placed a call but no one answered. "I will call him back later, and then you call me before eight tonight because I am leaving for Paris. I hope we can get this worked out before I am gone."

There was more small talk, but it was just that, small words to soothe large fears. We took a few photos to remember this most memorable occasion. Annick graciously offered Cuban cigars that her father had been given by Fidel Castro. De Bry happily took one and slid it into his pocket. A "victory cigar" he called it, for when the permits were signed.

But it was a glum walk back through the broken streets of Tana to the hotel, where we had nothing to do but wait to make the 8 P.M. call. When it came, it did not paint a positive picture of the days ahead.

It was the Malagasy ambassador who was now refusing to let us have excavation permits, Annick told us. Apart from her belief that power had gone to his head, she said there was something else going on that she could not put her finger on. "For some reason the ambassador does not like you," she told de Bry.

So what about the excavation permits? "Now you must deal with the local ministers. To get these papers signed could take days."

"You will help us, won't you?" asked de Bry.

"I will talk to you from the road," she said. "I am leaving for Paris now. Good luck."

It was a morose dinner that night at the Tana Plaza's excellent faux-French restaurant. Even an appetizer of pastry-wrapped zebu and a main course of rabbit osso bucco could not raise the spirits of the team.

For the time being, Plan A looked to be terminal. Now we were headed for Plan B—whatever that was.

NONE OF US COULD sleep that night. Jet lag and anticipation kept us from going to our rooms, so we sat on a rooftop patio at the hotel and drank beer until 3 A.M. There were long silences as we watched two teenagers in souped-up cars race on the wide Avenue of the Liberation. They would race for a quarter mile or so, turn around, and race back. This went on for hours. We talked until they returned, and then we waited until they roared off before talking again.

24

"MOVED AND SEDUCED"

KIDD HAD FELT ALONE, BETRAYED, AND deceived many times in his spotted career. But on May 8, 1701, the date of his preliminary hearing to examine the grand-jury charges against him, he felt all of those things at once. Surrounded by five lawyers for the Crown and facing the testimony of several of his most resentful men, Kidd was looking at charges that could end his life.

Standing in a courtroom at Old Bailey before Sir Salathiel Lovell, the recorder of London, Kidd listened in discomfort as the charges against him were read to the seventeen-man grand jury. The indictments included one for the murder of William Moore and three more for piracy. According to the court record, the grand jury withdrew to private chambers and returned with an indictment against him for murder and another against him and nine crew members of the *Adventure Galley* for piracy.

Proceedings like this usually moved fast, with the defendant declaring his guilt or innocence so the court could move on to the trial phase. Kidd, however, was different. Before pleading his guilt or innocence, he held up his hand and requested council. His request shocked Sir Lovell.

Recorder: What would you have counsel for?

Captain Kidd: My lord, I have some matter of law relating to the indictment, and I desire counsel to speak to it.

Clerk of the Court: What matter of law can you have?

Clerk of Arraigns: How does he know what it is he is charged with? I have not told him.

Sir Lovell: Captain Kidd, do you know what you mean by matters of law?

Captain Kidd: I know what I mean; I desire to put off my trial as long as I can till I can get my evidence ready.

Sir Lovell: William Kidd, you have best mention the matter of law you would insist upon.

Captain Kidd: I desire your lordship's favour; I desire that Dr. Oldish and Mr. Lemmon may be heard regarding my case.

Oldish and Lemmon were two lawyers who accompanied Kidd to the hearing in hopes of being allowed to represent him prior to charges being formally filed.

Clerk of Arraigns: What can he have counsel for before he has pleaded?

Sir Lovell: William Kidd, the court tell you that what you have to say shall be heard when you have pleaded to your indictment. If you plead to it, you may, if you will, assign matter of law, if you have any; but then you must let the court know what you would insist on.

Captain Kidd: I beg your lordship's patience till I can procure my papers. I had a couple of French passes, which I must make use of in order to make my justification.

KIDD HAD BEEN AUTHORIZED to attack French shipping. Lord Bellomont had forwarded to London confiscated certificates that proved that two of the vessels Kidd had taken had been sailing under French authority and consequently were legal prize. Kidd's hope was that presentation of these passes would short-circuit the Admiralty proceedings and put his case into a different court where conviction might be much less certain.

When he told Sir Lovell that he had last seen the passes in the possession of Lord Bellomont, the recorder declared that there was no reason to put off the trial and pressed Kidd to plead.

Captain Kidd: I beg your lordships I may have counsel admitted, and that my trial may be put off; I am not really prepared for it.

Sir Lovell: Nor never will, if you can help it.

Captain Kidd: If your lordships permit those papers to be read, they will justify me. I desire my counsel may be heard.

Mr. Coniers [lawyer for the Crown]: We admit no counsel for him.

Captain Kidd: I cannot plead until I have those papers that I insisted upon.

Mr. Lemmon: He ought to have his papers delivered to him, because they are very material for his defence. He has endeavoured to have them, but could not get them.

Mr. Coniers: You are not to appear for anyone till he pleads, and that the court assigns you for his counsel.

Sir Lovell: They would only put off the trial.

Mr. Coniers: He must plead to the indictment.

Sir Lovell: The course of courts is that when you have pleaded, the matter of trial is next; if you can then show there is cause to put off your trial, you may; but now the matter is to plead.

STILL KIDD REFUSED TO plead, demanding that he be given a chance to gather a legal defense to exonerate himself against at least some of the charges. Sir Lovell insisted that he plead his guilt or innocence. Only then, he said, could Kidd possibly have legal representation. Still Kidd refused to plead. Even though refusal meant an almost certain death penalty, it also assured that the possessions of his relatives would not be confiscated. Still the court pressed him for a plea.

Sir Lovell: You are accessory to your own death if you do not plead. We cannot enter into the evidence unless you plead.

Captain Kidd: If you will give me a little time to find my papers I will plead.

Clerk of Arraigns: There is no reason to give you time; will you plead or not?

Mr. Coniers: Be pleased to acquaint him with the danger he stands in by not pleading. Whatever he says, nothing can avail him till he pleads.

Sir Lovell: If you say guilty, there is an end to it; but if you say not guilty, the court can examine into the fact.

Clerk of Arraigns: William Kidd, art thou guilty or not guilty?

Captain Kidd: Not guilty.

Clerk of Arraigns: How wilt thou be tried?

Captain Kidd: By God and my country.

Clerk of Arraigns: God send thee a good deliverance.

KIDD AGAIN REQUESTED A delay of trial, but Sir Lovell now insisted that such a decision was not one he could make. "The judges will be here by and by, and you may move the court then; we are only to prepare for your trial," said Sir Lovell. "We do not deny your motion; but when the court is full they will consider of the reasons you have to offer."

Left to stew about his rights to representation and his bleak future, Kidd waited impatiently in an anteroom until the judges took their seats in court. Then the captain was led back into the courtroom and seated in the dock.

Told to stand and hold up his hand, Kidd remained sitting. The clerk of arraigns read the indictment against him.

The jurors of our Sovereign Lord the King do, upon their oath, present: That William Kidd, late of London, mariner, not having the fear of God before his eyes, but being moved and seduced by the instigation of the devil, on the thirtieth day of October, in the ninth year of the reign of our Sovereign Lord, William the Third, by the grace of God, of England, Scotland, France and Ireland, King, Defender of the Faith, and so forth, by force and arms upon the high seas, near the coast of Malabar, in the East Indies, and within the jurisdiction of the Admiralty of England, in a certain ship, called the *Adventure Galley* (whereof the said William Kidd then was commander), then and there being, feloniously, voluntarily, and of his malice aforethought, then and there did make an assault in and upon one William Moore, in the peace of God and of our said Sovereign Lord the King, to wit, then and there being, and to the ship aforesaid, called the *Adventure Galley*, then and there belonging; and that the aforesaid William Kidd, with a certain wooden bucket, bound with iron hoops, of the value of eight pence, which he the said William Kidd then and there had and held in his right hand, did violently strike the aforesaid William Moore in and upon the right part of the head of him the said William Moore, then and there upon the high sea, in the ship aforesaid, and within the jurisdiction of the Admiralty of the England aforesaid, given the said William Moore then and there with the bucket aforesaid, in and upon the aforesaid right part of the head of him, the said William Moore, one mortal bruise; of which mortal bruise the aforesaid William Moore, from the said thirtieth day of October in the ninth year aforesaid, until the one and thirtieth day of the said month of October, in the year aforesaid, upon the high seas aforesaid, in the ship aforesaid, and within the jurisdiction of the Admiralty aforesaid, did languish, and languishing did live; upon which one and thirtieth day of October, in the ninth year aforesaid, the aforesaid William Moore, upon the high sea aforesaid, near the aforesaid coast of Malabar, in the East Indies aforesaid, in the ship aforesaid, called the *Adventure Galley*, and within the jurisdiction of the Admiralty of England aforesaid, did die; and so the jurors aforesaid, upon their oath aforesaid, do say, that the aforesaid William Kidd feloniously, voluntarily, and of his malice aforethought, did kill and murder the aforesaid William Moore upon the high sea aforesaid, and within the jurisdiction of the Admiralty of England aforesaid, in manner and form aforesaid, against the peace of our said Sovereign Lord the King, his Crown and dignity.

How sayst thou, William Kidd, art thou guilty of this murder whereof thou standest indicted, or not guilty?

Captain Kidd: Not guilty.
Clerk of Arraigns: How wilt thou be tried?
Captain Kidd: By God and by my country.
Clerk of Arraigns: God send thee a good deliverance.

Kidd once again begged to have attorneys represent him. This time Justice Powell acquiesced, *if,* he said, there was truly a matter of law to be discussed. If not, Kidd must be tried.

The two attorneys, Oldish and Lemmon, moved in to make Kidd's case.

Dr. Oldish: My Lord, it is very fit his trial should be delayed for some time, because he wants some papers very necessary for his defence. It is very true, he is charged with piracies in several ships; but they had French passes when the seizure was made. Now, if there were French passes, it was a lawful seizure.
Mr. Justice Powell: Have you those passes?
Captain Kidd: They were taken from me by Lord Bellomont, and these passes would be my defence.
Dr. Oldish: If those ships that he took had French passes, there was just cause for seizure, and it will excuse him from piracy.
Captain Kidd: The passes were seized by my Lord Bellomont; that we will prove as clear as the day.
Mr. Lemmon: My Lord, I desire one word as to this circumstance; he was doing his King and country service, instead of being a pirate; for in this very ship there was a French pass, and it was shown to Mr. Davis,

and carried to my Lord Bellomont, and he made a seizure of it. And there was a letter writ to testify it, which was produced before the Parliament; and that letter has been transmitted from hand to hand, so that we cannot at present come by it. There are several other letters and papers that we cannot get; and therefore we desire the trial may be put off till we can procure them.

At this point the solicitor general, Sir John Hawles, became irritated. He declared that Kidd and his attorneys had had as much as two weeks to prepare for the trial. Dr. Oldish denied this, declaring that the money to wage a defense for Kidd had not been delivered to him until only the night before. It was true, declared Kidd. "I had no money nor friends to prepare for my trial till last night."

Finally the solicitor general offered a compromise. He would order Edward Davis to be brought from his cell at Newgate Prison. Since he had reportedly seen the French passes offered by the captured ships, he could take part as a witness for the defense in the piracy trial. In the meantime, the trial for the murder of William Moore should begin.

"[We should begin with that trial because] there is no pretense of want of witness or papers," he said spitefully.

And so it was. The murder trial of Captain William Kidd was to begin immediately.

25

BATTLE OF THE
FULL MOON

THAT EVENING WE SAT DOWN ON THE hotel porch overlooking the Avenue of the Liberation to work out a battle plan. Although we still weren't sure just what the battle was about, we thought that one minister or another might want us to purchase a "special permit," that Third World code for a bribe.

John de Bry and I had decided to stay in Tana to talk face-to-face with the powers that be on Monday. When de Bry spoke to Annick the next day in France, she agreed that it might be a good idea to stay in Tana another day or two. She would be back in Madagascar on Monday and looked forward to helping us.

De Bry's most recent conversation with Annick had me hoping that things would clear up quickly, and I thought it best to send the crew on to the island. That way they could unpack the gear and prepare to excavate once we received the permits. I was reasonably sure that by Monday the permits would be in hand.

The next day, sitting in a park near the hotel, I was approached by a man who was whistling in perfect pitch. He was dressed in shabby clothes and sweating hard, like someone in the grip of malaria's unpleasant symptoms. When he finished, he spoke in reasonably good English.

"What kind of stones are you interested in?" he asked.

I knew right away what he was talking about. Madagascar is noted for its precious and semiprecious stones. Everything from deep green emeralds to rich purple amethysts can be found in her mountains. Lately there had been a gemstone war in the southern part of the country. People had been fighting over small plots of land that may have contained gemstones and had created an environment similar to the Wild West, where much gold-prospector blood had been spilled fighting over mining rights.

And now one of the miners was standing in front of me, trying to ply his wares. Or was he? Maybe he was just a street hustler trying to unload worthless stones. I really didn't know. But for some reason I was feeling lucky.

"I would like a high-grade sapphire," I said as the man sweated.

"Wait here," he said, walking away. "I will bring you one that I found myself."

He returned shortly with a huge blue stone, about the size of both of my thumbs. It was beautiful, but for all I knew about stones it could have been colored glass. Still, I decided to take a chance. "It's nice," I said. "How much?"

"Two thousand dollars," he said without blinking.

We bartered and then bartered more. Finally we agreed on $160 American.

Early the next morning I took the stone to a jeweler whom I had done business with on our previous expeditions and asked him what it was worth.

"Oh, my goodness," he said, looking at the stone through his jeweler's loop. "This is worth quite a bit in your country, maybe over ten thousand dollars. You have made a very good purchase."

I was excited, not so much about the purchase I'd made as about the luck I hoped it foretold.

That afternoon we had a meeting with the president's diplomatic advisor, who placed a telephone call to Annick. She was in Tamatave with her father, she said, talking low into her mouthpiece. "In fact, I am sitting next to my father right now," she said. "I have spoken to him and I can promise action soon."

An hour later at 3 P.M., we called back. She needed to talk to the minister of justice, but she could not get him to return her calls. "Do not worry," she said. "There are no problems. You will get your permit, no problem."

The next day it started all over again.

"I have spoken to the minister of justice," Annick said, sounding happy at being able to present some progress. "There are procedures to be followed. The minister's secretary will get back to you with the papers you have to fill out. But don't worry, everything is fine."

Everything was not fine, not as far as de Bry was concerned. He had already filled out papers for the permits and did not understand why there were more. He decided to bluff. "I think it's over," he said to Annick. "If we don't get those permits soon, the Discovery Channel is going to cut us off."

"Oh, my, my, my," Annick said and laughed. "What is going on? It must be a full moon. I have talked to fifteen people today and everyone is feeling down. Don't change your plans because of a little paperwork. It will be delivered to you tonight. Finish it and then you can go to Sainte-Marie and finish your work. I will even meet you there."

We went back to the hotel and waited for the justice minister's secretary to show up with the papers. While we waited, de Bry called Limby Maharavo, the director of the office of the minister of culture whom we had dealt with so effectively on the past two expeditions. He seemed surprised about what was going on. "We have built a good professional relationship, so I have no idea what is happening," said Limby. "It is over my head."

We waited. Six o'clock came, then seven, and still no papers. At ten de Bry went to the lobby and searched the front desk himself, thinking the papers had been brought in and forgotten. At eleven the desk clerk called and said that a woman had just dropped off the papers.

De Bry came to my room stunned. What had been delivered were three large packages of forms that had to be manually filled out in quadruplicate. These were forms that required birth certificates, pictures, passport numbers, even proof that no one associated with the expedition had a criminal record, all of which we had provided them before we arrived for the expedition. De Bry was furious.

HE STAYED UP ALL night and filled out the documents to the best of his ability. In the morning we delivered them to the ministry and then prepared for our trip to the airport.

At this point another strange incident took place that once again filled us with false hope. As we were waiting by the curb for a taxi, a black Mercedes pulled up and the driver rolled down the window. "Mr. Clifford?" he asked.

"Yes?"

"I am sorry I am late. Please get in and I will take you to the airport."

De Bry and I looked at each other and smiled. This was obviously a private car for a very important person. Without saying anything we both thought the same thing, that one of the ministers was trying to make up for his mistake and had provided a car to take us in comfort to the airport.

Relieved to be in an air-conditioned environment and suddenly feeling better about the events of the last forty-eight hours, we sat back in the leather seat and looked out at Tana at high noon.

"This is a good sign," said de Bry. "When they let you have their car for any reason it's good news."

As we approached the airport, the driver suddenly took a turn toward a private airfield where a corporate jet was revving up on the runway.

"Excuse me," I said. "We're on the Air Madagascar flight. It's at the main airport. We aren't supposed to be here."

The driver stopped and looked at a piece of paper. "It says you are supposed to fly on a private plane," he said.

"Let me see that," said de Bry. He looked at the driver's orders and began laughing. "This car isn't for you, Barry," he said. "It's for someone named Ryan Clifford. We got picked up by someone else's car."

We all laughed, even the driver, who turned around and dropped us at the front door to the airport terminal.

"That was our first real good luck this whole trip," de Bry said.

I hoped it wasn't the last.

AFTER BREAKFAST ON SAINTE-MARIE the next day, I decided to do as much work on the site as our permits would allow. They clearly stated that we could film the site, which I took to mean that we could do anything but excavate. Since the Discovery camera crew wasn't coming until we got our permits in order, I decided to do a side-scan sonar survey of the harbor.

Such a survey would yield a picture of the harbor that would resemble an X ray done in strips. It would provide a clear image of every rock on the harbor's bottom, so clear in fact that we could differentiate between outcroppings of rocks, junk thrown into the harbor, and actual shipwreck sites.

We loaded my side-scan sonar equipment into Maximo Felice's boat. Wes Spiegel had rigged a spar off the side of the boat that allowed us to dangle the sonar Fish into the water. Once we pushed away from the dock, Spiegel lowered the Fish into the water until an image of the bottom came into view on Charlie Burnham's screen. Then we began to slowly "mow the lawn," trying to cover every inch of an ocean's surface.

The images that came up on the screen were so amazing that for the moment we forgot all of our problems. Standing out in relief against the harbor's black bottom were several piles of stone that were obviously the remains of old ships. Next to some were objects that looked like cannons and anchors.

WES SPIEGEL MANS THE SONAR, WHICH PROVIDES CHARLIE BURNHAM WITH A VIEW OF THE PIRATE-SHIP GRAVEYARD IN THE HARBOR. *Paul Perry*

"It looks like a pirate graveyard down there," said Burnham.

We went back and forth across the harbor, using the lighthouse on the point as our farthest boundary. Just beyond the point I noticed a boat anchored with several men on board. Some wore diving gear. "Who is that?" I asked Felice.

"That is another dive team from the United States," he said. "It's the team of Richard Swete. They're looking for an American ship, too."

Dick Swete—that was the archaeologist who was leading the expedition to search for the *Serapis,* the ship John Paul Jones had taken in his epic 1779 battle with the British off the coast of England. Through a strange chain of events, the ship became a private commercial ship and arrived at Île Sainte-Marie in 1781, where she burned up when her liquor stores were accidentally ignited. The ship went down right by what is now the harbor's lighthouse point, where we had seen the boat full of men.

I recalled de Bry's report of his courtesy call to Swete after our first expedition. Although Swete hadn't sounded thrilled that our crew would be working on Île Sainte-Marie, the conversation had been more or less amicable. But apparently something had been left unsaid by Swete. Felice reported now that since Swete's recent return to the island, he had been saying bad things about me. "He does not seem to like you very much and has told many people that he thinks you are here to take his shipwreck," Felice said.

I looked at the men on the boat and could see no one that I recognized. *If he just stays out of my way,* I thought, *I'll stay out of his.*

I had no idea that our collision course had been laid long before.

26

"Not Designedly Done"

Within a few hours of his preliminary hearing, Kidd was placed on trial for
the murder of William Moore. In an age of
fast trials, this one was especially speedy,
leading many to speculate that there were
members of the government who wanted
Kidd out of the way as soon as possible.

After Kidd was sworn in by the clerk of arraigns, three lawyers for
the Crown stood at the bar and offered their opening speeches against
the man accused of murder.

First up was Mr. Knapp, who read from the indictment and set the
ground rules for the jury. " 'That William Kidd, on the 30th of October,
on the high sea, on the coast of Malabar, did assault one William Moore,
onboard a ship called the *Adventure,* whereof William Kidd was captain,
struck him with a wooden bucket, hooped with iron, on the side of the
head near the right ear, and that of this bruise he died the next day, and
so that he has murdered the same person.' To this indictment he
pleaded not guilty; if we prove him guilty, you must find him so."

The solicitor general, Sir John Hawles, then offered his twopence.
"My lord, and gentlemen of the jury, we will prove this as particular as
can be, that William Kidd was captain of the ship, and that William
Moore was under him in the ship, and without any provocation he gave
him this blow whereof he died."

Those few words were followed by words from a third prosecutor, Mr. Coniers. "My lord, it will appear to be a most barbarous fact, to murder a man in this manner; for the man gave him no manner of provocation. This William Moore was a gunner in the ship, and this William Kidd abused him, and called him a 'lousy dog'; and upon a civil answer he took this bucket and knocked him on the head, whereof he died the next day."

The opening arguments were straightforward and over in a matter of minutes. Then, without a rebuttal by Kidd, Joseph Palmer was called to the stand. The crewman of the *Adventure Galley* told of meeting the merchant ship *Loyal Captain* on the coast of Malabar in the East Indies. There had been a desire by many of the men to take the *Loyal Captain* even though she sailed from Holland, a country friendly to England, said Palmer. In Kidd's mind, Moore was one of these men. Whether this was true or not, Kidd was simmering when he passed Moore.

Palmer: When Captain Kidd came on the deck and walked by this Moore . . . he came to him he said, "Which way could you have put me in a way to take this ship, and been clear?" "Sir," replied William Moore, "I never spoke such a word, nor ever thought such a thing." Upon which Captain Kidd called him a "lousy dog."

Then William Moore said, "If I am a lousy dog, you have made me so; you have brought me to ruin and many more."

Upon him saying this, Captain Kidd cried, "Have I ruined you, ye dog?" And took a bucket bound with iron hoops, and struck him on the right side of the head, of which he died the next day.

Moore said, "Farewell, farewell, Captain Kidd has given me my last."

It was time for Captain Kidd to cross-examine the witness. His case for killing Moore had been that the gunner was plotting a mutiny to turn the *Adventure Galley* into a pirate ship. It was thin gruel from which to make a case, but it was all that Kidd had. As Palmer sat in the witness box, his captain stood before him and peppered him with questions.

Sea captains were powerful in the lives of people like Palmer. To have to go against what one of them said—especially one as mean as Kidd—must have filled Palmer's stomach with fear.

"What was this Moore doing when this thing happened?" demanded Kidd.

"He was grinding a chisel," said Palmer.

"Was there any other ship?" Kidd asked.

"Yes, a Dutch ship," said Palmer.

"This ship was a league from us," Kidd said, "and some of the men would have taken her, and I would not consent to it, and this Moore said, I always hindered them making their fortunes. Was that not the reason I struck him? Was there a mutiny on board?"

"No," insisted Palmer. "You chased this Dutchman, and in the way took a Malabar boat, and chased this ship all the whole night; and they showed their colours, and you put up your colours."

"This is nothing to the point," insisted Kidd. "Was there no mutiny aboard?"

Palmer shook his head. "There was no mutiny," he said. "All was quiet."

Kidd pressed his point. "Was there not a mutiny, because they would go and take that Dutchman?"

"No, none at all," said Palmer, holding firm.

By now the jury must have been somewhat confused because one of its members addressed the witness from the jury box to ask why Captain Kidd had struck Moore.

Palmer replied: "A fortnight before [Moore was murdered] we met with the *Loyal Captain,* of which Captain Hoar [*sic*] was commander, and he came onboard Captain Kidd's ship, and Captain Kidd went onboard his, and then Captain Kidd let this ship go."

Apparently feeling that he was getting nowhere with this witness, Kidd desperately addressed the court in his own defense. "My lord, I was in the cabin, and hearing a noise, I came out," he said to Justice

Powell. "And William Moore said, 'You ruin us, because you will not consent to take Captain Hoar's ship. I will put Captain Kidd in a way to take this ship and come off fairly."

Hawles then rose and asked Palmer a single question: "Do you know of any other provocation to strike him besides those words?" he asked.

"No," declared Palmer.

THE NEXT WITNESS AGAINST Kidd was Robert Bradenham, the surgeon on the *Adventure Galley*. He told prosecutor Coniers that he was not present when Moore was struck but was sent for soon after the incident.

What he found was Moore lying on the deck, a small but fatal fracture denting his skull. Moore was still conscious but fading fast. As the surgeon examined the wound, the gunner moaned his last words. "Farewell, farewell," he said. "Captain Kidd has given me my last blow."

It was two months later, said Bradinham, that Kidd made a telling comment about the killing of Moore as well as the other trespasses he was on trial for. While talking to Bradinham during a reflective mood, Kidd cryptically referred to powerful friends in England who could get him off the hook for any of his misdeeds.

The surgeon recounted the words of Kidd: " 'I do not care so much for the death of my gunner, as for other passages of my voyage, for I have good friends in England that will bring me off for that.' "

In cross-examining his surgeon, Kidd had only one question: "I ask you whether you knew of any difference between this gunner and me before this happened?" he asked.

"I knew of no difference between you at all," Bradinham declared.

NOW IT WAS KIDD'S TURN to present the evidence for the defense. The captain rose and stated that he had evidence to prove he had not committed murder.

"My lord, I will tell you what the case was," he said, addressing the court. "I was coming up within a league of the Dutchman, and some of my men were making a mutiny about taking her, and my gunner told the people he could put the captain in a way to take the ship, and be safe.

"Says I, 'How will you do that?' The gunner answered: 'We will get the captain and men aboard. We will go aboard the ship, and plunder her, and we will have it under their hands that we did not take her.'

"Says I, 'This is Judas-like; I dare not do such a thing.' Says he, 'We may do it, we are beggars already.' 'Why,' says I, 'may we take this ship because we are poor?'

"Upon that a mutiny arose; so I took up a bucket, and just throwed it at him, and said, 'you are a rogue to make such a motion.' This I can prove, my lord."

As his first witness, Kidd called Abel Owens, a seaman from the *Adventure*. Owens insisted that there had indeed been a mutiny among the men who wanted to rob the Dutch ship. When Kidd found out about it, said Owens, he threatened to banish the sailors if they carried it out. " 'You will take the Dutchman, you are the strongest, you may do what you please,' " Owens said, offering his account of what Kidd had told the mutineers. " 'If you take her, you may take her; but if you go from aboard, you shall never come aboard again.' "

Next up for the defense was Richard Barlicorn, Kidd's cabin boy. The young Barlicorn denied that a mutiny had taken place, although he told Kidd, "There were many of the men would have gone with arms, and taken that ship without your consent."

But when asked if Moore died as a result of the blow from Kidd's bucket, Barlicorn insisted that the gunner died of a lingering illness. "William Moore lay sick a great while before this blow was given," said Barlicorn. "And the doctor said, when he visited him, that this blow was not the cause of his death."

Dr. Bradenham was recalled and reexamined by the solicitor general. He denied vociferously that it was illness and not the blow from Kidd's bucket that caused Moore's death.

Still, as the examination of Barlicorn continued, he continued to insist that "the doctor said he did not die of that blow."

"What did he die of?" asked Kidd.

"I cannot tell," said Barlicorn. "He had been sick before; we had many sick men aboard."

Kidd further attempted to discredit his ship's surgeon. As Bradenham looked on in disbelief, Kidd asked Barlicorn if the ship's doctor had been a part of the planned mutiny.

"If anything was to be, he was as forward as anyone," said the cabin boy. This was certainly true, since Bradenham was among the most vocal of those who defected to Culliford's gang.

The lord chief baron, a lawyer named Ward, could see where Kidd was going with this line of questioning. He spoke up and put a stop to it immediately.

"Captain Kidd, you are tried for the death of this Moore," he said. "Now why do you ask this question? What do you infer from hence? You will not infer that if he was a mutineer it was lawful for you to kill Moore."

In fact it was considered lawful for a captain to kill a crewman who was engaged in mutiny or piracy. Staggered by Ward's seemingly casual dismissal of the law, Kidd changed his line of questioning, which caused his meager defense to go flat. He had Barlicorn offer the theory once again that Moore was killed as a result of a simmering argument about Kidd's refusal to rob the *Loyal Captain*. Barlicorn did his best to convince the judges and jury that the death was a result of the righteous Kidd's battle against evil mutineers.

Kidd pursued that line in his closing arguments. Standing before the bar, Justice Powell asked Kidd if he had more to ask Barlicorn or any more witnesses to call.

"I could call all of them to testify the same thing," said Kidd. "But I will not trouble you to call any more."

"Have you anymore to say for yourself?" asked the lord chief baron.

"I have no more to say," said Kidd. "But I had all the provocation in the world given me; I had no design to kill him. I had no malice or spleen against him. . . . It was not designedly done, but in my passion, for which I am heartily sorry."

AFTER A RECAP OF the case by the prosecution, the jury retired to the jury room for deliberation. It took them barely an hour to make a decision. When they returned to the jury box, the clerk of arraigns asked them if they were ready to offer their verdict.

"Yes," answered the foreman, a man named Omnes.

"William Kidd, hold up thy hand," said the clerk. "Look upon the prisoner. Is he guilty of the murder whereof he stands indicted, or not guilty?"

"Guilty," said Omnes.

Dazed by the verdict, Kidd felt himself surrounded by the bailiffs.

"Look to him, keeper," said the clerk.

With a commanding tug, Kidd was escorted from the room and returned to his prison cell.

He would be hanged for murder.

27

A SON OF A PIRATE

WE HAD BEEN ON THE ISLAND ALMOST a week and had slipped into a predictable routine. We got up at 7 A.M. to eat a long breakfast, always hoping that this would be our last day without an excavation permit. Then we sat and planned the rest of the day, which would mostly involve sorting, cleaning, and testing gear. There would be smaller projects like rigging up an airlift or testing Charlie Burnham's tunnel cam, a tiny video camera that could be inserted through a pipe to examine the tunnels we had found on Pirate Island. It was common to see Bob Paine in the equipment locker reshuffling the gear he had sorted through the day before, or to catch Wes Spiegel testing radios or carrying a heavy case full of electronic equipment into Burnham's tech shack.

Stephanie de Bry, John's daughter and an archaeologist, could be found going over a checklist of equipment and procedures with Cathrine Harker and Layne Hedrick, a graduate of the noted School of Nautical Archaeology at Texas A&M. It would be their job to keep track of all of the artifacts once they were brought to the surface.

My son, Brandon, and Jeff Denholm carried their surfboards down to the lagoon, where they paddled over a mile out to the breakers and surfed with sharks over some of the sharpest coral any of us had ever seen.

I would make the rounds, trying to keep everyone's mood up.

STEPHANIE DE BRY STANDS FACE-TO-FACE WITH A SPIDER, ONE OF HUNDREDS OF EXOTIC INSECTS TO BE FOUND IN MADAGASCAR. *Paul Perry*

Eventually, though, I would borrow one of the hotel's mountain bikes and ride through the hills as hard as I could to burn off my frustration. Sometimes Margot and I would ride down to the bay and talk to Gregory, who was standing guard over the little bit of equipment that we still had on the shore near the wreck site.

"How is it going today, Mr. Barry?" Gregory would usually ask, grinning with his nearly toothless smile.

"Very slow," would be my stock reply.

"Yes," he would say. "The government here is like that."

JUST WHEN WE THOUGHT things couldn't get any worse, they did. It was October 20 and I was on the dining room porch reading the transcript from Kidd's trial when Fifou Mayer approached.

"Excuse me, Barry," he said nervously. "There are policemen here to see you."

Something in my stomach dropped as I stood up. A visit from the police in a Third World country is never good news. "What do they want?" I asked Fifou.

"I don't know," he said, then lowered his voice to a whisper. "But let me give you some advice. Do not offer a bribe, because they will think you have something to hide. Bribes have to go to the top. These little people have no power."

"What if they ask for radios or cameras or something like that?" I asked.

"Give them nothing," insisted Fifou. "In fact, don't even mention radios or radar or sonar or anything they may have seen in James Bond films. It will make them suspicious."

I nodded and strode as confidently as possible into the dark-paneled lobby. The entourage looked anything but suspicious. In fact they didn't even look up when I walked in. Fifou had brought them a plate of toast and a side of butter that might have amounted to more than a pound. Now they were covering the toast with thick slabs of butter and jamming it into their mouths, hurrying to gobble it down before the man without a uniform stood and spoke.

"Good morning," said the pleasant young man, sticking out his hand. "I am the supraprefect of the town. This is the chief of police . . . the head of the gendarmerie . . ."

I shook hands with each in turn, smiling outwardly but filled with dread. This meeting was not about bread and butter. A passage from *Treasure Island* flashed through my mind. *Look out for squalls when you find it. . . .*

"We have come to tell you to stop any activities until the permits have arrived," said the supraprefect.

I was stunned. "Our passports give us permission to film," I said. "We can't excavate, but at least we can film."

"Not anymore," said the pleasant young man. "The regional president has decided to stop you until these matters get cleared up."

"What matters?" I asked.

He looked perplexed. "I am not entirely sure," he said. "Can I please see your permit to film?"

I handed him my passport and watched as he read the directive from the Washington ambassador. Fifou brought out more toast and butter, and the five officials dove back into the offering as if I wasn't even there.

"It was nice to have you here today," I said as convincingly as possible. All five nodded and continued to eat. I walked away.

When the toast was gone and the officials had left, de Bry, Fifou, and I tried to figure out exactly why we were being prevented from excavating—and now filming—on the site. Without permits to excavate and film, the Discovery Channel wouldn't even send the film crew.

To get to the bottom of the problem, Fifou made calls to the justice minister's son, and de Bry once again placed a call to Annick. The calls went on nonstop for the rest of the day. Slowly the real situation at hand emerged: we were being badmouthed by archaeologist Richard Swete not just on the island but in influential places.

"Why would Richard Swete be saying bad things about us?" I asked.

It was a mystery. Although de Bry had spoken to Swete after the first expedition, he knew little about him, and Layne Hedrick, the archaeologist hired by de Bry to help excavate our site, knew the name but nothing more. Finally I asked Kinkor in Cape Cod to do an Internet search, and Perry called a reporter in Arizona who agreed to ask an acquaintance of mine, adventure writer Clive Cussler, who had worked with Swete. With only Swete's name to go on, the two compiled a sparse dossier that shed no light on the mystery.

A Vietnam veteran, Swete had lost his left leg below the knee to a land mine and was discharged from the army. After getting his master's degree in nautical archaeology from Texas A&M and a doctorate in history from the College of William and Mary in Williamsburg, Virginia, he had helped Clive Cussler search the James River in Virginia for a Civil War naval vessel, the *Florida*.

"I barely remember Dick," Cussler said. "The last time I saw him was almost twenty years ago."

Kinkor's research turned up only that Swete had attempted to find John Paul Jones's *Bonhomme Richard*.

I was stumped and angry. A person I didn't know was trying to stop my work for seemingly no reason. We might be on the same island, but we were not at all interested in the same shipwreck. Indeed, our sonar survey of the harbor showed that there were enough shipwrecks for several archaeological teams. There was no logical answer to why Swete would try to stop my expedition. "Maybe he just doesn't like you," de Bry offered.

That night we found out just how much he didn't like me. Shortly before dinner, the police and the supraprefect arrived again, this time accompanied by an army colonel and his attaché. They drove up to the front of the hotel in a Citroën so small and strange-looking that it resembled a clown car in a circus.

The passenger door opened and the colonel stumbled out. At first his instability seemed to be simply the result of the car's cramped quarters. Then, as he staggered toward the hotel, negotiating the front porch with difficulty, it became clear that we were dealing with a drunk soldier.

Someone knocked on my bungalow door. "Hey, Barry. There's a drunk army officer here to see you."

By the time I arrived in the lobby, Fifou had covered the table with a spread of beer and cracker bread. The way to a Third World official's heart was clearly through his stomach.

As we shook hands all around, the tension in the room was palpable. I noticed that none of the uniformed visitors were wearing guns, which was encouraging, since everyone but the supraprefect seemed to have been drinking liberally before arriving. I waited for them to sit, and then the three of us—myself, de Bry, and Paul Perry—sat, too.

"I am a criminal specialist," the colonel declared, awaiting our response as de Bry translated.

"Have I done something criminal?" I joked.

"Not at all. I am here to assess the quality of your people and to look at you. Normally I would have sent only a deputy. But because of the importance of the affair, I wanted to come on my own."

The colonel's name was Patrice Boto. He had trained with the United States Coast Guard in some capacity, and proudly wore a pin on his chest to prove it. He also wore a number of other pins signifying a wide variety of accomplishments. He was here, he said, to make sure we were "doing the right thing."

I must have looked puzzled when he said that because he repeated it and added further comment: "I want to make sure you are following the law. That all of your equipment came in legally and is legal equipment. I am not here because you are white people on our island. It doesn't matter to me if you are white, red, or blue or anything else. I just want to know what you are up to." He launched into a lengthy monologue, talking about his childhood on the island, his life as a military man, and his suspicions that we had somehow broken the law in a way that he couldn't quite put a finger on.

Apparently everyone else was as confused as I was. The chief of police took his beer and went outside, and the supraprefect, who was sober, was struggling mightily to explain what exactly it was they were doing there. "We just want to know what it is you are really wanting to do here," he said. "We have heard things, and we want to know what is true."

Before I could ask what he had heard, the colonel began another monologue. To decide what they were going to do, he had to see all of our equipment and have it demonstrated so he would know how it worked.

"Plus," he said, dropping the final bomb. "I need to have all of your passports, and you must all get on the boat to Tamatave for questioning at the police station."

When de Bry finished translating, we looked at one another in stunned silence. I didn't know whether to laugh or become angry. The entire scene resembled a skit from *Saturday Night Live.* The joke, however, would be on us if we submitted to his demands.

"John, I think it's time to call Annick," I said.

De Bry fired off some words of French that immediately jolted the colonel out of his stupor. Before the colonel could say anything more, de Bry stormed out of the lobby and into Fifou's office, which was within earshot of where we were meeting.

Somewhere in Paris, Annick was just climbing out of a car on her way into a discothèque when her silver cell phone rang. She stopped on the sidewalk and put the receiver to her ear. Over the sound of Paris traffic and the disco's throbbing beat Annick could hear John recounting forcefully and in uncomfortable detail the meeting with the drunken colonel.

"You must get his name to me immediately," said Annick. "Tell him I will deal with this when I return!" She spoke loud enough that the colonel could hear the anger in her voice. It had the expected effect. He looked nervously around the room and then began to smile. His beard grew only on his neck and gave him the look of a very nervous billy goat as he set his beer down on the thick wooden table.

"You did not get it," he said. "What I said was a joke. The part about taking you to Tamatave for questioning especially was a joke."

The tables had turned now, and the colonel was suddenly becoming cooperative. It was then that we learned the true reason for his visit.

"An American archaeologist went to the regional president and told him you are not to be trusted," the colonel said to me. "He told the president that you know where a lot of treasure is and that you plan to take it."

"We don't sell artifacts," I insisted, struggling to cope with my anger. "We are here to make a film about a famous American pirate, not to find treasure."

"Perhaps," said the colonel. "The president does not know for sure.

But he has become tired of hearing this from the other archaeologist. He has ordered both of your projects to be shut down until the issue gets settled."

"So you were just sent here to hassle us, is that right, *Corporal*?" said de Bry, hinting at the level of demotion he hoped the colonel would receive when Annick returned from France.

The colonel shrugged and took another swig of beer. "No, I have been sent here to prevent violence," he said clearly. "The regional president tells me that he does not want white people fighting on his island."

THE NEXT DAY FIFOU confirmed Richard Swete's involvement through a friend in the ministry of justice. It seemed as though Swete started rumors as soon as he heard we were coming to the island. Through a friend who was the curator at a museum in Tana, Swete had declared that we were crooks disguised as a film crew. He contacted the Washington ambassador, saying the same thing. Soon our path was blocked every way we turned. Now the regional president had shut down both of our teams, which was the only prudent thing he could do from his point of view.

I rounded up de Bry and Perry and asked them to do me a favor. To avoid further squabbling or even a confrontation, I asked them to visit Richard Swete to see if there was a way of resolving this issue.

"Know thy enemy," said de Bry, quoting Shakespeare.

"Right," Perry agreed. "To have a grievance is to have a purpose in life."

Together they disappeared down the dusty road that would take them to the Chez Pierrot, a spare and rugged hotel occupied by Swete and his expedition of seven. A truck was leaving as de Bry and Perry walked up the path to the hotel. The people in the truck bed seemed alarmed that anyone from the "rival" expedition would wander into their camp. One fellow banged the top of the cab with the flat of his hand and jumped out when the driver stopped. He looked like a car-

bon copy of Glen Campbell, the country musician. He was joined by a Malagasy man who immediately introduced himself as an archaeologist from the university in Tana. "Are you with the Barry Clifford expedition?" asked the Glen Campbell look-alike, an archaeologist with a faint Australian accent.

"We are," said de Bry. "We've come to see Dick Swete."

The archaeologists led the pair around the hotel office and into the café. "Wait here for a minute," said the Australian. "I'll go get Dick."

A few moments later an extremely thin man emerged from one of the bungalows and limped across the open yard. Judging by his disheveled look, he had obviously been sleeping. He greeted de Bry and Perry with a cordial handshake, sat down, and lit an unfiltered cigarette. The hotel owner brought a chilled Three Horse Beer out of the kitchen and set it in front of Swete. Over the next two hours, de Bry and Perry watched in awe as Swete emptied six of the potent twenty-four-ounce beers. The cigarette count went over fifteen before they lost track.

"Looks like we've all been shut down because of your expedition," said Swete with a sly smile.

His declaration took de Bry and Perry by surprise. "Well, we think it's the other way around," said de Bry.

Swete clouded the air with another puff and then made his mission statement: "I hate Barry Clifford, and I'll do anything I can to stop him."

"Even if you stop yourself?" asked Perry.

"I am only stopped temporarily," he said. "The last team here will be the one that gets the excavation permits. And I can wait you out because I have a government pension. For me, the eagle shits every month."

"What happened between you and Barry?" asked de Bry.

"The *General Arnold,*" said Swete.

"The traitor?" asked Perry.

"No, not the traitor. The ship," said Swete. "Clifford stole a ship from me in 1977."

The ship he was talking about was an American privateer named the *General Arnold.*

On Christmas Eve 1778, with a crew of 105 men, the ship filled with water during a heavy gale and sank in about ten feet of water and sand off Plymouth Harbor in Massachusetts. Water and snow swept the vessel, trapping the survivors on the quarterdeck. Huddled together, the sailors died of exposure. By the time rescuers could reach the ship, seventy bodies had frozen together. Ten of the frozen dead were eventually separated and buried in coffins. The remaining sixty were frozen solid and had to be buried in a mass grave near the town of Plymouth. It was one of the East Coast's most appalling shipwrecks.

I had searched for this ship at the suggestion of Larry Geller, the curator of the Plymouth Pilgrim Hall Museum. Using aerial surveillance I had been able to find the remains of a ship in the area where the *Arnold* sank.

Shortly after I found it, a promoter named Charles Sanderson claimed to have found it first. He wanted to raise what was left of the ship and turn it into a museum. He hired an archaeologist and a legal team, and they tried to beat me out of the claim. Swete was the archaeologist who worked for Sanderson.

A court had ruled in my favor, upholding my claim, but Swete apparently still felt as though his people were entitled to the *Arnold.* He was still harboring his grudge after all these years.

Perry knew the entire story of the *Arnold,* including an embarrassing portion that Swete chose not to tell. After I fought so hard to keep my claim to the shipwreck it turned out not to be the Revolutionary War ship at all. The real *General Arnold* had been refloated.

When I discovered this, I gave up my claim. Sanderson and his group, with Swete as the head archaeologist, began their own excavation, promising extraordinary relics from the early years of our country's fight for independence. That was when I went to the press and announced the reason I had renounced the claim.

"Dick, this all happened twenty-three years ago," said Perry to Swete. "Why don't you just let it go so we can all work?"

"Simple," said Swete. "In archaeology, we fight so hard because the stakes are so small."

As it turned out, however, the stakes weren't so small after all. Eventually the other issue came out: the presence of the Discovery Channel. When he initially found the *Serapis,* Swete had approached the Discovery Channel about funding his expedition by making a documentary film of the search. The producers liked the idea, said Swete, but they were already dealing with someone else who had found a pirate wreck on the very same island. "When they told me it was Barry Clifford, I could feel my blood pressure rising," said Swete. "First he stole a ship from me and now the chance for funding. As far as I'm concerned, I can't get in his way enough."

There seemed little left to say, so de Bry and Perry stood to leave.

"I've been admiring that ring on your hand," Perry said.

It was gold and embossed with a double-headed phoenix, the mythical bird that is resurrected from burning embers. It looked similar to the symbol found on the Austrian porcelain, but Swete said his represented a pirate, not a royal family. "That is the crest of William Rogers," said Swete, referring to the nineteenth-century English pirate who operated off the Gulf Coast of the United States. "Somewhere back there I am related to him. This ring has been handed down to me."

He thought a moment, swigged some beer, and grinned. "I guess that makes me a son of a pirate in a way," he said proudly.

"So we're in a fight with a pirate," I said to de Bry and Perry.

"That's right," said de Bry. "A bona fide relative of a famous pirate. And one with a wooden leg at that."

I remembered the *General Arnold* incident very well, and Swete himself was coming into vague focus now from the days when I saw Chuck Sanderson in court. Swete was obviously an angry man who had purposefully lied to stand in my way. I wanted to get to the bottom of this, once and for all.

When I checked with the Discovery Channel I could find no one who remembered ever talking to Richard Swete, let alone expressing interest in his expedition. What was going on?

"You've got to let it go," said my Zen-like son, Brandon. "Anyone who carries a grudge for two decades has a lot of emotional luggage."

I agreed with Brandon, and I had to let it go for another reason, too: we were running out of time. We'd been here nearly three weeks, and the Discovery Channel was getting restless. David Conover was calling me daily to remind me that he would not be able to bring a camera crew from the States until we had a new permit certified by the appropriate government officials.

Rather than focus on Swete, I had to direct all of my energies toward convincing the government that Swete was not telling the truth. I had to persuade them that we should be allowed to carry on as planned by excavating the site and proving once and for all whether this was the *Adventure Galley* or another pirate ship.

Through de Bry's considerable talents as a negotiator and Fifou's connections with the minister of justice, the regional president became convinced that we were a legitimate expedition involved in the legitimate act of making a documentary film. But reversing his stop-work order was impossible, he said.

"Why is that?" asked de Bry.

"Because I can only put orders like this in place," he said. "To have them erased, you must get the authorization of three cabinet ministers: the ministers of justice, transportation, and culture."

I broke into a cold sweat.

It was nearing Halloween, a major holiday in Madagascar. It was unlikely that any cabinet ministers would be in their offices, and, if they were, that they would make a decision on an issue like this quickly. This was a hot potato. It would have to cool on the desk of a politician for a few days before being resolved.

"What should we do?" asked de Bry.

"Everything we can," I said. "Tomorrow is too late. Hit them with everything. Today's the day. Call Annick."

In a few minutes de Bry was pleading our case on the telephone to Annick. In less than an hour Annick called de Bry back and said that she was meeting all of the ministers at the airport that evening and would see if they would issue a temporary permit before Halloween, only two days away.

"Why at the airport?" asked de Bry.

"Oh, I don't know," she said. "I think some of them are going away for a few days."

My heart sank even further. The producers at the Discovery Channel had been very patient up to this point, but I knew that if the project dragged into another month, they would likely pull the plug. I did not want to come home empty-handed. I needed to know the truth about that ship's identity.

28

The Greatest
and the Worst of All

THERE WERE THREE MORE TRIALS IN store for Kidd, all for robbery and piracy.

The charges in the first trial declared that Kidd "did piratically and feloniously set upon, board, break, and enter a certain merchant ship, called the *Quedagh Merchant* . . . and there piratically and feloniously did make an assault in and upon certain mariners."

The second trial a few hours later charged that Kidd did "piratically and feloniously set upon, board, break and enter a certain ship called a Moorish ship . . . did steal, take and carry away 100 pound weight of coffee, of the value of £5 of lawful money of England, 60 pound weight of pepper of the value of £3 of lawful money of England, 1 cwt. of myrrh of the value of £5 of lawful money of England, and 20 pieces of Arabian gold of the value of £8 of lawful money of England, the goods, chattels and moneys of certain persons (to the jurors aforesaid unknown)."

The third trial had two indictments, the first for robbing a "Moorish ketch" of "£50 of lawful money of England; thirty tubs of sugar-candy, of the value of £15 of lawful money of England; and ten bales of tobacco, of the value of £10 of lawful money of England, the goods and chattels of certain persons . . . then and there upon the high sea."

The second indictment declared that Kidd did "steal, take and carry away two chests of opium, of the value of £40 of lawful money of England, 80 bags of rice, of the value of £12 of lawful money of England; one ton of bees wax, of the value of £10 of lawful money of England; a half ton of iron, of the value of £4 of the lawful money of England" plus the goods and chattel of persons unknown to the jury.

In the course of the trials, the entire events of the previous five years were laid before the jury. Witnesses spoke of the promises of wealth that were made to them before they signed on with Kidd in both Plymouth and New York. They told of the mysterious illness that swept the ship and decimated the crew and described Kidd's dark moods that led him to kill William Moore and threaten other members of the crew.

They talked of the taking of several ships and about how the *Adventure Galley* filled with water and nearly came apart on the high seas.

Joseph Palmer, the damning witness in Kidd's murder trial, now took over in the witness seat. Near Karwar, Palmer said, Kidd fought a ship and took her. "He took a bale of pepper, and some myrrh to use instead of pitch," said Palmer. "He took about 60 pound weight of pepper and about 30 pound weight of myrrh. I cannot say what was the value of the Arabian gold he took. I did not see any take out then, but a pretty while afterwards . . . he gave every mess two pieces of Arabian gold. This was about ten or twelve days after the ship was taken. The pepper was divided among the messes, and all the prisoners had their share of it. Several of the men on board this Moorish ship were hoisted up and drubbed with a naked cutlass."

Another story from Palmer accused Kidd of robbing the natives of the island of Malabar after a battle at sea with a Portuguese merchant vessel. According to Palmer, Kidd also "took one of the natives, and bound him to a tree, and shot him to death; I saw the houses on fire."

The witnesses for the prosecution recounted a chain of piracies that led all the way to Île Sainte-Marie in Madagascar, the place where Kidd was supposed to arrest as many pirates as he could find. Arrests were never made, said Palmer.

"You have produced letters . . . that empowered you to take pirates; why did you not take Culliford?" asked Justice Powell.

"A great many of my men were gone," insisted Kidd.

"But you presented him with great guns and swore you would not meddle with them," said Justice Powell.

"When the question was put, 'Are you come to take us and hang us?' you answered, 'I will fry in hell before I will do you any harm,' " said the lord chief baron.

"That is only what these witnesses say," said Kidd.

Exasperated, the lord chief baron leaned toward Kidd. "These things press very hard upon you," he said. "We ought to let you know what is observed that you may make your defence as well as you can."

KIDD DECLARED THAT THE witnesses against him were telling "a thousand lies." He offered little in the way of defense. He had taken the Moorish ship *Rupparell* (later renamed *November*), but he had done so legally, since his commission allowed him to seize French ships. Both the *Rupparell* and the *Quedagh Merchant* had been flying French colors and carrying a French pass.

"I have many papers in my defence," declared Kidd. "If I could have them."

"What papers were they?" asked the lord chief baron.

"My French passes," said Kidd.

"Where are they?" asked the lord chief baron.

"My Lord Bellomont had them," declared Kidd.

"If you had anything of disability upon you to make your defence, you should have objected to it at the beginning of your trial," said the lord chief baron. "What you mean by it now I cannot tell."

The French passes taken from the *Rupparell* and the *Quedagh Merchant* represented the only evidence that may have saved Kidd. He was not granted access to this evidence, even though these passes can be found to this day in the Public Record Office in London. The captain was hung out to dry.

KIDD STOOD IN THE courtroom as the jury returned to render its verdict. Once again Foreman Omnes had the duty of speaking for the jury.

"William Kidd, hold up thy hand," demanded the clerk of arraigns. "Look upon the prisoner. How say you? Is William Kidd guilty of the piracy and robbery whereof he stands indicted . . . or not guilty?"

"Guilty," said the foreman.

"William Kidd, hold up thy hand," said the clerk of arraigns. "What canst thou say for thyself? Thou has been indicted for several piracies, and robberies, and murder, and hereupon hast been convicted. What hast thou to say for thyself who thou shouldst not die according to law?"

"I have nothing to say, but that I have been sworn against by perjured and wicked people," said Kidd. "My lord, it is a very hard sentence. For my part, I am the innocentest person of them all, only I have been sworn against by perjured persons."

As the bailiffs led him from the courtroom, Kidd must surely have felt the weight of the day on his shoulders. Three more capital crimes had been charged against him; three more reasons why he would now be heading for the gallows.

29

THE TRICK-OR-TREAT
SHOW

 IN ITS OWN STRANGE WAY, HALLOW-
een in Madagascar is a celebration of resur-
rection. The locals believe that during the
days surrounding that October 31 holiday
the dead arise, take stock of the affairs of
the living, and return to their place of rest
beneath the ground.

In a way we were hoping for the power of this holiday to somehow
revive our comatose expedition. We prayed it would infect the three
ministers who were withholding our permit to excavate, so that, in an
ironic twist, we could then slip into our neoprene diving outfits and
provide a costume party for the townsfolk of Ambodifotatra. Our
prayers were almost answered. The ministers had a special meeting
regarding our situation, but in the end it yielded nothing.

The crew, meanwhile, waited. They had now seen and done everything
this island had to offer. They had been diving and snorkeling in the crys-
tal-clear Indian Ocean, swum with whales cruising toward the feeding
grounds of the Antarctic, taken four-wheel adventures to remote beaches,
seen the exotic wild animals that have earned Madagascar the title of "the
Land That Time Forgot." John de Bry had even been bitten by an angry
lemur, an unfortunate encounter that had earned him five stitches.

The only thing the crew hadn't done yet was the job they came to do, excavate the remains of the *Adventure Galley*. There was nothing left for them to do but wait.

IT WAS NOW NOVEMBER 2. Paul Perry and de Bry had just flown back from Tana, where they'd met with the minister of culture, Fredo Betsimifira. A handsome man with the lean, hungry look of a middleweight boxer, Betsimifira had granted the meeting but refused to stand up or shake hands when the two were escorted into his massive office in the Tana government complex. With them was Fabrice Digiovanni from Les Lézards de Tana. In the minister's waiting room, he delivered his analysis of African politics.

"The problem is this," he said, filling the hotel lobby with a cloud of smoke from a Gauloises cigarette. "You just have not learned to deal with the African way of doing things. When you say to them 'Time is money,' they cannot understand what you mean. And they do not like to be pushed."

The minister's frightening countenance made Perry and de Bry think that they had pushed things too far. They sat down across from the impeccably dressed official. He stared—glared really—at the three people across the desk. Perry would later say that he felt like a mouse being played with by a big cat.

After a few words of greeting, the minister cut to the chase. "You have spoken to Annick lately?" he asked.

De Bry began telling our story but was cut off by an impatient wave of the minister's hand. He then picked up a cell phone and punched in a number. In a moment he was greeting Annick.

His mood changed and he became playful as he talked to her about her recent holiday. Then he asked if she was helping us and fell silent as he listened to her response. When he got off the phone, the minister was a changed man.

"I will talk to the transportation minister," he said, "but I feel certain that I will be able to get you a permit that will allow you to dig in your

WHILE WAITING FOR PERMITS, MARGOT NICOL-HATHAWAY SEEKS OUT
LEMURS AND OTHER WILDLIFE, AND BOB PAINE SWIMS WITH A PASSING
WHALE. *John de Bry/Paul Perry*

part of the harbor. But tell Mr. Clifford that he must stay in his part of the harbor, and away from Mr. Swete. We do not want problems with white people on the island of Sainte-Marie."

De Bry asked if it was possible to get the permit now, before they returned to the island.

"No," said the minister emphatically. "I will try to get it to you tomorrow."

But tomorrow never came. The minister did not produce a permit before Perry and de Bry boarded the plane the next day for Sainte-Marie, nor did he fax it to our hotel. When de Bry called his private number, he did not answer. Once again the minister of culture had disappeared.

Once again we called everyone who we thought could help. The minister of justice, his son, the minister of culture, Annick. My anger and disbelief rose with each call. Nothing seemed to be changing. "This has all been the act of one person," I said to de Bry. "One guy lies about you, and everything gets screwed up."

"Yeah," said de Bry. "And in this country it takes a long time to unscrew them."

We ended up talking again to Fabrice Digiovanni, who was in Tana, acting as an intermediary for Limby Maharavo. It was Friday night now, the end of the workweek. Maharavo, through Digiovanni, was saying that he needed the money to complete our papers. But when pressed to promise that we could go to work the next day, he balked. "I cannot promise that," he said cheerfully. "I can only promise that I will start the papers and get them to you sometime."

Forget it, I told him. "No permits, no money," I told Digiovanni, who, I felt, was stonewalling for money.

A few minutes later Digiovanni called back. Maharavo had left for the weekend, he said. There would be no chance to get the permits before next week.

My expedition was over.

We were scheduled to return on November 5, only two days away.

FRENCH GUIDES FABRICE DIGIOVANNI *(LEFT)* AND GILLES GAUTIER. *John de Bry*

Some of the crew members had families and jobs, and staying as long as they had already represented a hardship. To ask them to stay longer was not an option. Besides, we had no permits and none appeared to be coming in the near future. We had been stopped by lies and mired in bureaucracy. Our third return to Treasure Island was for naught.

Dinner that night was a grim affair. In fact most of the dinners had been grim affairs. They took place in the dark, in a candlelit, open-air dining room that was set with fine china and native waiters squirming uncomfortably in formal wear.

I looked around the table at the crew, and without a word we toasted. I was sad and angry. By my best estimates I had spent untold amounts of money, time, and energy on this expedition. After finding the coins I was convinced that there was plenty of evidence down there that could tell us which ship we were dealing with. I had thought of staying beyond our planned departure date many times in the last couple of weeks, and now, in the final hours, I knew I had to. I could not be beaten by lies. *The one who is here last gets the permit,* Swete had said.

There at dinner that night I decided that I would be the one to get those permits. If it was a war that Swete wanted, then he could have it.

"I'm not going to leave," I said to de Bry, who had sat down next to me. "I'm going to wait here until Christmas if I have to. I am not going to get beaten by Dick Swete."

"Okay," said de Bry, in a singsong voice that told me he already knew what was coming next.

"Can you stay, too?" I asked him, holding my breath. Without an archaeologist I couldn't and wouldn't excavate the site, even if the permits did come through.

"Okay," said de Bry, that same singsong tone in his voice.

"That means you have to stay for the duration," I said.

"Okay," he said.

And that was it. We would remain after everyone left and wait for the permits. And if we got them, we would excavate.

"I wonder what Christmas is like here," said de Bry dryly.

I honestly hoped that we didn't have to find out.

30

TWICE TO THE GALLOWS

DESPITE ALL THE EVIDENCE TO THE contrary, Captain William Kidd could not accept that he would soon be hanged for murder and piracy. He told the Reverend Paul Lorrain, the ordinary of Newgate who was assigned to provide religious comfort in his final days, that he would soon be sprung from jail, reprieved by one of his many highly placed friends in London.

"I found him unwilling to confess the crime he was convicted of, or declare anything, otherwise than that he had been a great offender and lived without any due consideration either of God's mercies or judgements or of his wonderful works which had often been set before him," wrote Lorrain of his futile attempts to get Kidd to confess his sins before the Almighty. "That he never remembered to return Him thanks for the many great deliverancies he had received from Him nor called himself to account for what he had done."

Thanks to the keeper of Newgate Prison, Kidd did have many visitors. He had friends and relatives in London, and the keeper made extra money by admitting family members and curious citizens who came to see notorious prisoners.

Among his visitors were old friends such as Sarah Hawkins and her husband, and the widows of his sailing master and quartermaster. Several officers of the ship that brought him to justice from Boston also visited the now infamous rogue.

But no one came to free him.

And still Kidd held out for deliverance. On a daily basis the Reverend Paul Lorrain would preach to Kidd and the other members of his crew who were going to be hanged in only a matter of days. Every day Lorrain's sermon was the same: those who failed to repent would "go away into everlasting punishment." And every day Kidd refused to ask for forgiveness or to provide the confession of his sins that Lorrain thought was so necessary for his salvation. As Lorrain later wrote, Kidd remained adamant to the end that he would be rescued, "vainly flattering himself with hopes of a reprieve."

It was a reprieve that never came.

Still Lorrain persisted. In his twenty-year career, he amassed an estimated five thousand pounds by providing religious comfort to convicts and writing accounts of their repentance on the scaffold. He certainly knew how to persist, as he did with Kidd. He dogged Kidd daily for his confession, reading passages from the Bible that offered him a free pass into heaven if only he would confess his sins. Whether Kidd was delusional and truly believed that he would be rescued at the last moment or was simply arrogant is not known. As far as the minister was concerned, Kidd had chosen the path to hell.

On the morning of May 23, Kidd's last day on earth, the Reverend Paul Lorrain took the unrepentant pirate to the Newgate Prison chapel, where he "administered further admonitions of repentance." When that failed he brought him there again in the afternoon. Finally, to his great joy, Kidd relented.

"I was afraid the hardness of Capt. Kidd's heart was still unmelted," wrote Lorrain. "I therefore applied myself with particular exhortations to him and laid the judgements of God against impenitent and hardened sinners, as well as his tender mercies to those that were true and sincere penitents, very plain before him. To all of which he readily assented and said that he truly repented of his sins and forgave all the world, and I was in good hopes he did so."

The hour of his execution was now near, and Lorrain went ahead of Kidd to Execution Dock at Wapping where a crowd had gathered in anticipation of a good day's entertainment, the hanging of five pirates, including Darby Mullins and the famous Captain William Kidd.

Back at the chapel, things took a strange turn for the pirate captain. Someone, perhaps the jailer, had slipped Kidd a large bottle of rum, and finally realizing that he was destined for the gallows, Kidd began to drink heavily.

The Captain William Kidd who arrived at the gallows that late afternoon of May 23, 1701, was a different Kidd from the one Lorrain had left in the chapel. He rode in an open cart and was wrapped in chain secured by a padlock. Ahead of him walked officers of the Admiralty carrying a silver oar that signified their authority to take Kidd's life. Between the officers of the Admiralty and Kidd were the Admiralty marshal in his carriage, accompanied by two city marshals on horseback. Behind Kidd were a squadron of marshals and sheriffs, assigned to protect the prisoner from the thousands of people who had gathered on Commercial Road to see the pirate swing. The road was so packed with spectators that the procession could hardly reach the gallows. Hanging from the houses that lined the road were morbid sightseers. In the water were dozens of small boats crowded among ships, their riggings filled with sailors jockeying for a view of the wooden gallows that had been erected on the muddy shore near Globe Wharf.

No longer penitent, Kidd was now a drunk and surly man. He spat and swore at the spectators around him as they sang a street ballad called "Captain Kidd's Farewell to the Seas," which went:

Some thousands they will flock when we die, when we die,
Some thousands they will flock when we die,

Some thousands they will flock
To Execution Dock
Where we must stand the shock and we must die.

"I found to my unspeakable grief, when he was brought thither, that he was inflamed with drink, which had so discomposed his mind, that it was now in a very ill frame," wrote Lorrain. "I prayed for him and so did other worthy Divines that were present, to whom (as well as myself) the Captain appeared to be much out of order, and not so concerned and affected as he ought to have been."

Hundreds of people watched, "ribald and roistering" as one author of the day wrote. Two Frenchmen were hanged and then another pirate, John Eldredge, was granted a reprieve. Then it was Darby Mullins's turn to die. Lorrain had spent time with Mullins. The forty-year-old from Londonderry, Ireland, had survived a devastating earthquake in Jamaica, had survived the death of his wife in New York, had survived the "bloody flux" at Mohelia but was now certain he would not survive this day. He begged Lorrain for "pardon of God" and confessed that he had been a great sinner and had not served God as he should. "He wished he had not been such an offender," wrote Lorrain. "He had of late very much given himself up to swearing, cursing and profaning the Sabbath Day, which had deservedly brought this calamity upon him."

Mullins was escorted to the gallows and prepared for hanging. The noose was looped around his neck, and he was positioned over the trapdoor. Then he was dropped. The crowd cheered as he wriggled and slowly expired.

It was Kidd's turn.

Lorrain was desperate for Kidd to show a change of heart, especially since he was preparing a pamphlet of Kidd's confession, *Only True Account of the Dying Speeches of the Condemn'ed Pirates,* that he planned to sell to the general public the very next day. Lorrain's only problem was that Kidd was not being as contrite as he needed him to be. To make

CAPTAIN KIDD HUNG OVER THE RIVER THAMES FOR MONTHS, AN EXAMPLE TO ALL WHO MIGHT CONSIDER A LIFE OF PIRACY. *Artist unknown*

these pamphlets perfect, the prisoner had to confess his sins and ask forgiveness in the eyes of God. So far that had not happened, at least not to the extent Lorrain wanted.

Rather than declare himself a sinner and ask God for forgiveness, Kidd launched into another round of self-defense. As the crowd, officers, clergy, Admiralty officials, and journalists leaned toward him, Kidd began a desperate defense of his innocence. As diarist Narcissus Luttrell recorded it, Kidd

could hardly be brought to a charitable reconciliation with those persons, who were evidences against him, alleging that they deposed many things that were inconsistent with truth and that much of their evidence was by hearsay. In the general part of his discourse he seemed not only to reflect on them but on several others, who instead of being his friends as they professed, had traitorously been instrumental in his ruin! He further declared that as to the death of William Moore, his gunner, the blow that he gave him, it was in a passion, as being provoked by him to do so, but not with an intention to kill or murder him. Nay he was so far from bearing any malice against him, that he freely gave £200 for his ransom, and further said that all his sailors knew he always had a great love and respect for him, adding that if any one concerned in his tryal had acted contrary to the dictates of his or their own conscience he heartily forgave them and desired that God would do the like. . . .

He expressed abundance of sorrow for leaving his wife and children without having the opportunity of taking leave of them, they being inhabitants in New York. So that the thoughts of his wife's sorrow at the sad tidings of his shameful death was more occasion of grief to him than that of his own sad misfortunes. . . . He desired all seamen in general, more especially Captains in particular to take warning by his dismal unhappiness and shameful death and that they would avoid the means and occasions that brought him thereto, and also that they would act with more caution and prudence, both in their private and public affairs by sea and land, adding that this was a very fickle and faithless generation.

At the end of his speech, Kidd was helped up the gallows steps and placed in the hands of the hangman. The two were a match for each other on this warm Friday afternoon. Both were drunk and barely able to stand up. Finally the hangman looped the rope around Kidd's neck and set the trapdoor free. Kidd fell and his body jerked at the end of the rope. The crowd cheered.

Then a miracle: the rope snapped.

The dazed and drunk pirate landed in the mud beneath the gallows. Gasping for air, he looked around at the crowd of people, many of them roaring at the irony of the events playing out before them.

The Reverend Paul Lorrain saw his last chance. He raced to Kidd's side and once again tried to extract a confession. Kidd looked at the clergyman with great bewilderment. He had just been hanged, but now he lived. Perhaps this was the second chance he had so arrogantly insisted that he would receive all along, the chance to confess his crimes to God. As Lorrain wrote:

When he was brought up and tied again to the tree, I desired leave to go to him again, which was granted. Then I showed him the great mercy of God to him in giving him (unexpectedly) this further respite that so he might improve the few moments now so mercifully allotted to him in perfecting his faith and repentance.

Now I found him in much better temper than before. But as I was unwilling, and the Station also very incommodious and improper for me, to offer anything to him by way of question that might perhaps press him to embrace (before it was too late) the mercy of steadfast faith, true repentance, and perfect charity. Which now he did so fully and freely express, that I hope he was hearty and sincere in it, declaring openly that he repented with all his heart, and died in Christian love and charity with all the world. This he said as he was on the top of the ladder (the scaffold being now broken down) and myself half way on it, as close to him as I could, who, having again, for the last time, pray'd with him, left him, with a greater satisfaction than I had before that he was penitent.

Kidd was strung up again, and this time the rope held. To the grue-some cheers of the crowd, he wriggled at the end of the rope, doing what any of us would try to do, suck in another breath and hope for a second miracle.

What grim irony, that the view from those gallows was of the dock on the River Thames where his fateful voyage on the *Adventure Galley* had begun.

With the crowd roaring its approval, Kidd's life expired—and his legend was born.

As was the custom of the day, Kidd hung at Wapping until the tide washed over his body three times. Since the hanging had taken place at low tide, there was no need to move him for quite a while. Later his body was put into a cage called a gibbet and moved to Tilbury Point, farther down river. There, according to some sources, the body was suspended for several years, left as a warning against piracy to passing seamen.

Ballads sprang up in England well before his rotted body was removed from the gibbet, and in the colonies, too, where poems and plays were written that made him a folk hero. Here is an abbreviated version of one that circulated via broadsides. It was called "The Dying Words of Capt. Robert [*sic*] Kidd":

> *YOU captains brave and bold, hear our cries,*
> *You captains brave and bold hear our cries,*
> *You captains brave and bold, tho' you seem uncontrol'ed*
> *Don't for the sake of gold lose your souls, lose your souls,*
> *Don't for the sake of gold lose your souls.*

My name was Robert Kidd, when I sail'd, when I sail'd,
  My name was Robert Kidd, when I sail'd
My name was Robert Kidd, God's laws I did forbid,
  And so wickedly I did when I sail'd.

My parents taught me well, when I sail'd, when I sail'd,
  My parents taught me well when I sail'd,
My parents taught me well to shun the gates of hell,
  But against them I did rebel, when I sail'd.

I made a solemn vow, when I sail'd, when I sail'd,
  I made a solemn vow, when I sail'd.
I made a solemn vow, to God I would not bow,
  Nor myself one prayer allow, when I sail'd.

I murdered William Moore as I sail'd, as I sail'd,
  I murder'd William Moore as I sail'd;
I murdered William Moore, and I left him in his gore,
  Not many leagues from shore, as I sail'd.

Take warning now by me, for I must die, for I must die,
  Take warning now by me, for I must die;
Take warning now by me, and shun bad company,
  Lest you come to hell with me, for I must die;
  Lest you come to hell with me, for I must die.

31

"NO ONE SHOULD
DIE ALONE"

THE DAY, LIKE MANY ON THIS THIRD expedition, started with bad news. Fabrice Digiovanni called to say that he couldn't find any of the government ministers. They were all on holiday for Halloween.

I had come to expect phone calls like this, but they still left me depressed and frustrated. I rented a bicycle from the front desk and went riding along some of the island's fine wilderness trails with Brandon. It was November 4 now, and most of the crew was going to spend the day preparing for their return to the United States tomorrow.

We didn't get back until late in the day. As I rode into the hotel's sandy driveway, I could see Paul Perry and John de Bry climbing into a tiny panel truck. Perry was carrying our bulky medical kit. They both looked grim.

"What's going on?" I asked.

"It's Dick Swete," said Perry. "Maximo's wife just called from the hospital and said that he is dying. She asked us to bring all of our medical equipment to the hospital. She doesn't think he'll make it through the night."

I was totally taken aback. Swete had been a thorn in our side, the reason we could not work. His petty and vindictive slander had caused the government to shut down my expeditions. Now he was dying. Was it from a contagion that might affect my crew? Would we soon have need of all of our own medical supplies? What if we were incurring some kind of legal liability by being Good Samaritans? My mind raced. It didn't matter.

De Bry and Perry grabbed the medical equipment and left for the hospital, a trip that would take nearly an hour over eight miles of bad road. I had been to the hospital. I knew that Perry and de Bry were headed to a filthy facility, practically devoid of any medical equipment whatsoever. I remembered talking to Gregory, the native we hired to guard the expedition tent, about medical care. When I mentioned to him that a person with a serious problem could go to the hospital on the hill, he looked at me with genuine horror. "That is the kind of hospital you go to when you are going to die," he said.

This is not going to be an easy night, I told myself. I went back to my bungalow to change clothes and then found Fifou. "Let's go up to the hospital," I said. And we left.

WE ARRIVED AT THE hospital to find a scene of horror. Dick Swete was lying in a sweat- and urine-soaked bed, his head tipped back, his eyes wide open. Surrounding him were de Bry, Bob Paine, and Perry, assisting Claudia Allegra, Maximo Felice's wife, who had just recently become an M.D.

I walked around the bed and got close to Swete's face. His eyes seemed to widen slightly and his raspy breathing picked up, but beyond that there was little reaction. He was in a coma.

"Dick, it's me, Barry Clifford," I said, hoping he would be shocked to consciousness.

There was no response.

"He can't hear you," said Perry. "He's not really with us anymore."

The room was starting to fill up. One of the local doctors was there, looking as helpless as I had ever seen a doctor look. The supraprefect was there, as was the head of the gendarmerie. This time they were here to help.

In the corner of the room two local French divers were talking to de Bry about calling Swete's wife in Sacramento, California. De Bry recommended against it since we did not know Swete's status for sure, but they went outside and made the call anyway.

"Where's his crew?" I asked de Bry.

"They all left him," he said. "The owner of the hotel is here. She said that a couple of them got sick and went back to Texas. The others went to Tana last week to wait until the permits came. They told her they were tired of waiting."

Dr. Allegra, standing at the foot of the bed, had been the first doctor to be called by the locals and had been caring for him since. Using our medical equipment (the hospital had almost none), she had started a line to administer glucose and had catheterized him so his bladder would not become too full and burst. Beyond that there was little she could do.

"He will die without help," she said. She looked forlorn. "The owner of the hotel Chez Pierrot where he was staying told me that Swete became sick four days ago. There was blood on his pillow when she went to clean the room. She asked if he wanted to go to the mainland to see a doctor, and he refused. After complaining of illness for two more days, he decided to quit drinking beer, definitely cause for alarm, the owner said, because Dick Swete liked his beer. This morning he packed his bags and said he was going to move to a nicer hotel to get some rest. Then he lay down for a nap and never woke up. She checked him later and found him lying in a pool of blood and breathing with difficulty. That was when they drove him here to the hospital and contacted me."

By Allegra's tentative diagnosis, *esophageal varicies,* Swete's drinking had caught up with him by causing the arteries in his esophagus to rupture from liver damage. "In a good hospital I could probably save him," said Allegra. "But they don't have the right equipment here. They don't even have anesthesia. If we don't get him off the island and into a hospital, he will die."

But was it really a ruptured esophagus that was killing Swete? Could it be a tropical disease—perhaps a worm from the bay where we were diving—that had eaten a hole in his stomach? I suddenly became paranoid. Could my crew suffer the same fate?

I expressed my fears to Dr. Allegra, who assured me that no such flesh-boring worm was known to exist in these waters. This was cold comfort. David Conover, the Discovery Channel producer, had been medevaced on a previous occasion for an unknown ailment. What if this was something worse?

Darkness was descending, and rescue would become increasingly difficult, since the airport had no landing lights. "We have to get him out of here," I said.

I called the Divers' Alert Network, an excellent organization that offers worldwide medical evacuation to divers in distress. They were concerned, but once I told them where we were they said that no plane could possibly reach us before morning.

I hung up and called Bob Rhule at the American embasssy in Tana. Luckily he had given me his home telephone number when I had met with him after we had first arrived. When he answered I explained Dick Swete's condition and told him that we needed to get him off the island as soon as possible.

"It might take a while," he said. "It's against the law to fly airplanes at night in Madagascar without specific authorization from one of the cabinet ministers. But I'll get right on it."

After a few rounds of telephone calls by Rhule, Fifou, his father, and many other people, we received authorization from the minister of justice to bring an airplane in for Swete's evacuation. The idea was to have him picked up at the Sainte-Marie airport and flown to Réunion Island, where there was a French hospital with modern medical equipment.

In order to light the runway here on the island, Fifou's father contacted a number of people with cars, asking them to ring the tiny asphalt airstrip with their lights turned on so that the Cessna could see the landing zone. The embassy called back and said that the airplane would arrive in ninety minutes.

A large piece of plywood was placed in the panel truck to form a stable platform, and then a number of us picked up Swete's mattress with him on it and carried him to the truck. We slid him in as far as he

could go and elevated his head with pillows to prevent him from aspi-
rating any blood that he might vomit on the trip. His legs were hanging out the back of the tiny vehicle but were supported by the plywood.
On one side of him sat Dr. Allegra. On the other de Bry held the bag of
glucose high so it could continue to drain.

A motorcade formed ahead of Swete's truck to accompany him
down to the airport. At the head of the line was a tiny blue car from the
gendarmerie, followed by two other cars and a couple of motorcycles.

"We're going to take off now," I said to Swete, putting my hands on
his legs as I spoke. I had forgotten about his false leg, and the feel of the
hard prosthetic jolted me for a moment. Then a thought flashed into
my head: *This is not the first medevac for this guy.* When he was younger,
Swete had stepped on a land mine in Vietnam and had been evacuated
from a jungle much like he was about to be now. He had survived that
horrendous war wound, I told myself. Maybe he can survive again.

The motorcade lurched forward and started its painfully slow trip to
the airport. We tried to drive as cautiously as possible to keep from
traumatizing Swete, but the unplowed road made difficult going. The
roads were jagged with rocks and rutted with potholes that at times
had us scraping the truck's carriage. From our position behind Swete's
truck, we could see him getting quite a shaking.

The road crossed the causeway, next to the ships we were both looking for—the *Adventure Galley* for us, and the *Serapis* for him. On we
drove, by the discothèque, which was throbbing now with Saturday-
night fever. Several miles down the road we passed the Chez Pierrot,
where Swete had spent the last three weeks brooding about my presence on the island. Another mile of dark jungle and we were in front
of the Princesse Bora. The motorcade stopped.

"I am going to call the airport in Tana to make sure the plane has
left," said Fifou, running into his office. After a few minutes of waiting I
went into the hotel, too, only to find Fifou talking agitatedly into the
phone.

"The plane hasn't taken off yet," said de Bry, translating as Fifou spoke. Apparently they had not yet received the green light from the minister of justice.

Fifou called someone else in the government and finally got the airplane cleared for takeoff. It was now after 9 P.M., and it would be at least two hours until the rescue plane from Tana would arrive.

We drove onto the airport runway and parked on the tarmac. I had the driver turn the truck with the open back to the east so Swete could get the full benefit of the cool winds blowing in from the Indian Ocean. Dr. Allegra continued to check Swete's vital signs. They weren't good. He was gasping hard now for air, and his eyes were wide and desperate.

"I wish this had happened someplace else," said Dr. Allegra, her eyes filled with frustration. "We cannot give him a blood transfusion because there is no way to type blood here. And we cannot do surgery because there is no place to operate. At home I might be able to help him. But here . . ."

Her voice trailed off to keep from saying the obvious.

By now many curious locals had arrived at the airport to see what was going on. They gathered in groups and would sometimes drift over to the back of the truck and peek in to see if Swete was still alive. Some paid their respects, saying short prayers or waving as though they knew he was not going to come back.

"Let's elevate him again," said Dr. Allegra. As she and de Bry held him up to put another pillow under his head, blood began to foam from Swete's mouth. His eyes fixed on a distant spot, and a groan came from deep inside.

And that was it. Like a candle under a bell jar, Dick Swete's flame had expired.

DR. ALLEGRA CLOSED SWETE'S eyes and mouth and checked him for a heartbeat one more time. Then she began wrapping him with a sheet and saying a silent prayer. Several of my crew members were red-eyed. Some were gently sobbing. Although none of us really knew Swete, there was a sadness in the air that was impossible to choke back.

"Think of this guy's misfortune," said de Bry, summing up the irony of the moment as I gazed at Swete's shrouded body. "Here he is alone, halfway around the world, dying with strangers who happen to be people he hates."

Someone, perhaps it was Fifou, used his cell phone to call the pilot. He had been in the air only five minutes, so Fifou told him to return to the Tana airport and come back tomorrow. It was too risky to make a night landing just to pick up a dead body.

I used the telephone in the airport manager's office to call the American embassy and report the death of a United States citizen. Once again I spoke to Rhule, who told me that he would arrange to have the body picked up in the next two days. "Who's with him?" asked Rhule.

"He's alone," I said. "His crew had all abandoned him. We had to take care of him until he died."

"Someone has to make an inventory of his possessions to make sure nothing gets stolen," said Rhule. I promised to get it done.

Back at the truck, de Bry and Paine were shocked that I had promised to make an inventory of Swete's possessions, and they called over Fifou. He shook his head adamantly. "You can't do that, Barry," he said. "There is a tradition on this island of poisoning one's enemy. If you start looking through his luggage, people here will think you killed him to get his shipwreck, or maybe you know of a treasure map that he has in his bags. Already some people are suspicious."

Being a murder suspect held no appeal to me, especially on an island where superstition and tradition could take the place of justice. I deferred to Fifou. "So what should I do?" I asked.

"You should just walk away. He is dead now and no longer concerns you. Just walk away."

And that is what I did. I walked across the tarmac and back down the dark road that led to the hotel. In a few moments Paine passed us in the panel truck, watching over Swete's body in the back.

The native driver looked horrified. I found out later that the truck

EXPEDITION MEMBERS BOB PAINE, CATHRINE HARKER, AND
CHRIS MACORT STAND ON THE AIRPORT TARMAC MOMENTS
AFTER RICHARD SWETE'S DEATH IN THE TRUCK AT THE
LEFT. HIS LUGGAGE IS PACKED AND SITTING NEXT TO THE
TRUCK. *Paul Perry*

had to be shipped off the island and sold because none of the locals
would ride in a vehicle that had carried a dead man.

At the police station Swete's body was carried out of the truck and
laid on the warm cement. It stayed there for more than a day before
being transported to the mainland.

I was walking down the beach in front of the runway when the airplane bearing Swete's body lifted off on its short flight to the mainland. Upon the return of his body to the United States, the Centers for
Disease Control in Atlanta performed an autopsy. They concluded that
Swete had died of cerebral malaria, *P. falciparum,* the most deadly of
that disease's four strains. So bad was his case that the CDC asked the
family if they would donate his organs for research instead of cremat-
ing him as Swete had requested in his last will and testament.

Now I watched as the airplane climbed for altitude, quickly becoming a fleck of metal in the blue sky. Then I turned for the hotel and
walked slowly back.

32

THE UNREQUITED LEGACY

THE END OF KIDD WAS BY NO MEANS the end of the Kidd legend. Ballads and tales of piracy—both true and false—swirl around his infamous name to this day.

It was mostly Kidd himself who launched dozens of tales of concealed treasure when he asked to be freed from jail in Boston, promising Bellomont that he would retrieve thousands of pounds of gold that he claimed were still aboard the *Quedagh Merchant* in Hispaniola. The existence of this gold proved to be a myth, but that didn't keep more myths from brewing.

In Kidd's lodging in Boston, for instance, one thousand pounds worth of gold dust and ingots were found along with a bag of silver, leading some to believe that there was a lot more where that came from. It was even suspected that the buttons on Kidd's waistcoat were diamonds. Lord Bellomont personally checked out this rumor and proclaimed the buttons to be gold dabs topped off by low-grade "Bristol stones," a faux diamond of the day. He was wrong. They turned out to be the real thing.

Dark and incredible tales of Kidd at Gardiners Island sprang up. In fits of madness, Kidd is said to have scattered paper money around the house. One story has him dropping pearls into wine and drinking it. Others tell of him burying treasure in a swamp and showing the frightened Lord Gardiner where it was. "If it is not here when I return, I will kill you," Kidd is rumored to have told Gardiner.

Even Robert Livingston contributed to the legend of buried gold when he stated under oath that Kidd "had Forty pound weight in Gold which he hid and secured in some place betwixt this [Boston] and New York, not naming any particular place, which nobody could find but himself." He made this statement after an official inventory of Kidd's booty put the amount of gold at 1,111 ounces and silver at 2,353 ounces.

The rumors of concealed treasure from Kidd's exploits at sea must have weighed heavily on the ever-broke Bellomont. He hunted down Captain Gillam, who was ultimately jailed along with Kidd in Boston, and found a letter from Kidd ordering a Captain Paine to give Mrs. Kidd twenty-four ounces of gold "but to keep the rest until further notice." A further search of a residence where Gillam had stayed uncovered no gold. The home of Captain Thomas Paine, himself a retired pirate and a friend of the Kidd family, was searched. Even Mrs. Kidd was shaken down for the gold. Twice in ten days officials took everything of value from her home, including the family silverware. She protested the loss of the silverware to Bellomont, and he ordered it returned. Nothing was found.

According to the record, several meetings of the New York Council in July and August 1699 concerned themselves with the subject of Kidd's missing treasure. This led to the examination of several other people, including John Tuthill, the justice of the peace in Suffolk, who was accused of concealing Kidd's treasure; Carsten Luesten and Hendrick van der Head, two sailors who were believed to have taken some of Kidd's goods from Gardiners Island; a Captain Carter, Kidd's former quartermaster; a "little black man"; and Thomas "Whisking" Clarke, the coroner of New York.

"Whisking" Clarke actually had a portion of Kidd's treasure. A trunk of Kidd's that was found in Clarke's possession contained as much as twelve thousand pounds. Clarke told the earl of Bellomont to mind his

ONE OF THE MANY LEGENDS SURROUNDING CAPTAIN KIDD IS THAT HE BURIED HIS BIBLE IN A RENUNCIATION OF MAN AND GOD. EXACTLY WHERE HE BURIED IT IS NOT PART OF THE LEGEND. *Artist unknown*

own business, and Bellomont said that he was. He had Clarke arrested and delivered with his goods to the lieutenant governor of New York. How much money this arrest truly added to Bellomont's coffers is not known. It must not have been much, however, since Bellomont continued to search in vain for Kidd's riches.

In September 1699, Bellomont received a letter from Colonel Peleg Sanford, judge of the Admiralty Court in Rhode Island, saying that Captain Gillam was rumored to be in Boston. This was good news for Bellomont, since Gillam had returned from Madagascar with Kidd and was said to have gone ashore in Delaware with two chests.

Bellomont immediately dispatched a constable to make the rounds of all the taverns and inns of Boston in order to find Gillam. The constable located Gillam's horse tied outside the first tavern he went to. But when he stepped inside to make the arrest, the owner told him that Gillam had left just fifteen minutes earlier. Another constable was posted at the tavern in hopes that Gillam would return for his horse, but the veteran pirate was too wary for that. The horse stood there for days before being taken away by the constable.

A wanted poster was printed offering 208 pieces of eight for the capture of Gillam. Within two days a solid lead developed. Someone brought word that a Captain Knott might know where Gillam was. When Knott denied any knowledge of Gillam, Bellomont summoned his wife, who nervously told a different story. She said that a man named James Kelly had stayed with them for several days before leaving for Charlestown across the river. Bellomont once again turned to Knott to find out where Gillam had gone in Charlestown. Sheepishly he finally confessed that his nefarious friend was in a house of prostitution run by Francis Dole.

A group of deputies surrounded Dole's house and searched it. They found no Gillam.

Later in the afternoon, two of the deputies were searching the forest near Dole's house when Gillam came walking across a field. Suspecting nothing, he introduced himself to the surprised deputies and told them he was on his way back from "treating two young women some few miles off in the country." He was arrested and eventually hanged in London. The money was never found.

LORD BELLOMONT CONTINUED TO chase Kidd's treasure almost until the day he died, which was just months before Kidd was hanged. He never collected more than a few thousand pounds of the pirate's ill-gotten gain. Bellomont certainly rued the day he got involved with Kidd, and rued perhaps even more so his decision not to let Kidd go back to Hispaniola. Despite Kidd's silence about the whereabouts of his treasure, his adventures sparked a treasure mania that swept the New

England coast and lasts to this day. One of the first to note it was Benjamin Franklin, who wrote in 1729, "You can hardly walk half a mile out of the town on any side without observing several pits dug . . . there seems to be some peculiar charm in the conceit of finding money."

William Smith, who wrote about this treasure mania in his 1757 book, *A History of New York,* said that there was hardly a spot of coastline or an island that didn't bear the scars of the shovel from digging for pirate treasure. "Some credulous people have ruined themselves by these researches, and propagated a thousand idle fables current to this day among our country farmers."

People turned to witchcraft to find gold; they used verses from the Bible, divining rods, and dreams in their search for its location. Ghosts and other malingering spirits seen in remote areas were thought to be guarding treasure.

Kidd's treasure was everywhere. And nowhere.

For a while it was even believed that Kidd had sailed the *Adventure Galley* all the way to New York from Madagascar and scuttled it far up the Hudson River, where it was sunk with the treasure still on board. This legend gained such a strong foothold that in 1829 Abraham Thompson, a descendant of the Gardiners, purchased 100 acres of land that bordered the Hudson River. He planned to search for the treasure ship, which he felt was within 250 feet of his place on the shore. Using iron poles and boring augers, the team probed for the ship. Their efforts were in vain, and when one member of the expedition died in an accident, the search was stopped.

In 1844 it was renewed again, this time by another group. They had better luck. From the muck of the river the treasure hunters brought up a piece of timber and an old cannon. They decided to construct a cofferdam around the site and pump out the water until it was dry. A skeptical article in the New York *Journal of Commerce* told of their effort:

The good people are still at work around Captain Kidd's vessel. They have enclosed her in a thick wall or dam supposed to be watertight. They have a steam pump in operation throwing out the water, with which they make considerable impression.

The first half hour they lowered it four inches. At that rate they will soon be rioting in great masses of gold which have quietly reposed in her hold for more than a century, far beyond the covetous grasp of man. A shaft has been sunk on the side of the mountain opposite the ship, from which specimens of gold are said to be obtained.

This may be owing to the approximation of such an immense quantity in the hold of that ship.

Ultimately, no gold was ever found.

The organizers of this ill-fated venture produced a fund-raising pamphlet entitled *An Account of Some of the Traditions and Experiments respecting Captain Kidd's Piratical Vessel*, which speculated that there might be as much as ten million dollars' worth of treasure still under the mud and made Kidd out to be some kind of a dark superhero who used Satanic intervention to steal great amounts of treasure.

"His vessel was often seen streaking its way along the Sound everything set, while all other craft were double reefed or settled away," read the pamphlet. "Whenever he was closely pursued by the English men-of-war that were sent out for his capture, he would be rescued by the interposition of some violent story, or that the evil spirits themselves would come to his aid by some surprising manifestations." The pamphlet went on to summarize seven of the Captain Kidd treasure legends related to the Hudson River so that potential investors could get an idea of the "scarcely conceivable" amount of booty that was out there.

In 1846 a noted psychic named Mrs. Charles Chester produced a pamphlet entitled *The Wonderful Mesmeric Revelation* in which she went into a trance and saw Kidd's ship sunken up the Hudson River. She saw Kidd, whom she described as being "large, stout, of large chest, shoulders and neck, who had, furthermore, a Roman nose, piercing eyes, and a very broad head, whose personality was characterized by a great cautiousness, combativeness, and destructiveness, in fact, the *tout ensemble* of a bloodthirsty *filibustier.*" Of course, she also saw treasure, "chests filled with bars of solid gold; decayed shot bags spilling with heaps of precious stones including diamonds, gold watches like duck's eggs in a pond of water, a diamond necklace and the remains of a beautiful young lady."

Not all of the treasure hoaxes were of the metaphysical variety. In 1849 two boys found a bottle with a letter inside in Palmer, Massachusetts. Breaking it open, they discovered what looked like an old letter that was dated 1700 and signed Robert Kidd.

To John Bailey, Esq., New York.

Sir: I fear we are in a bad situation. We are taken for pirates and you must come to Boston as soon as you get this. . . . If I do not see you, I will tell you where my money is, for we have plenty of that if it will do us any good it is buried on Conant's Island in Boston harbor on the northwest corner of the Island in two chests containing from fifteen to twenty thousand pounds sterling in money jewels and Diamonds. They are buried about four feet deep with a flat stone on them and a pile of stone near by. There is no one that knows where it is but me now living as Dick Jones and I hid it when part of my men were in Boston and the rest asleep one night; it is about sixty rods up the side hill. . . . I want you to see Col. Slaughter and John Nichols, Esq. And James Bogard and Captain Houson and Edward Teach and all that can do me any good. . . . They think I have got the money buried down at Plymouth or down that way somewhere, they don't think it is so near to Boston. . . . Come quickly, if I am gone for England, secure money or diamonds and follow. It will buy a great many people. . . . Keep dark. . . . They keep me well here, this is from your friend.

Robert Kidd

When this letter was read to the townspeople of Palmer, a fight erupted between the fathers of the boys who had found it and the letter was put in a bank vault while the local magistrate decided on whom the rightful owner should be. Before a decision could be reached, one of the fathers sheepishly admitted to having written the letter and put it in the cave as a hoax.

And so it goes to this day, madness and fantasy inspired by dreams of gold. Although there may still be "Kidd treasure" buried somewhere, none has ever been found.

33

THE BROTHERHOOD
OF PIRATES

I STAYED IN MADAGASCAR ON THE outside chance that I would finally be given permits to excavate the wrecks, hoping that Dick Swete had been right when he said, "The last team here will get the permits." The needless war with Swete had cost us money, and plenty of it, as well as precious time we could have used exploring some of history's most interesting shipwrecks. And that was just our reckoning. The price Swete had paid was far higher.

And so I waited. I sent everyone except John de Bry and Jeff Denholm home to the United States.

On November 8 de Bry flew to Tana to meet with the minister of culture. He'd had several meetings with the stern young bureaucrat and was counting on another grueling conversation and more empty promises. But this time it was different. Fredo Betsimifira showed some genuine interest in the project and expressed regrets that "unforeseen circumstances" had held up the expedition. After a long conversation in the minister's office, de Bry was told to return in the morning. He expected nothing from the next day's meeting. Having been through this routine before, he figured that the minister would promise that the permit would be issued after he got the needed signatures from the other ministers.

But that was not the case. In the morning Betsimifira told de Bry that we would immediately be issued a "short permit," one that would allow us to excavate and film the site for a maximum of ten days. The permit would start the next day, November 10, and continue until the twentieth. He requested that we not return the relics to the wreck site as we had before. Instead he wanted us to keep them in a secure place on the island.

It was late in the afternoon when de Bry called me in Sainte-Marie with the news. He wouldn't be able to fly to the island until the next day, but when he did we would start excavation immediately. I telephoned David Conover, the Discovery Channel producer, and told him that we finally had a permit in hand. Then, deciding I needed another diver, I called Bob Paine in Boston. He was still recovering from jet lag after his long flight home but agreed to fly back with the film crew to help with the excavation.

Denholm and I worked late into the night preparing the equipment for the next day's dive. When we had everything together, we went to our bungalows to catch a few hours of sleep. That was when I had the strangest dream of my life.

I found myself sitting in an ancient coliseum next to a man with a long white beard and dressed in a robe made of broadcloth. Around us was a group of robed men chanting a mantra. I followed them down a narrow passageway to a tall, eight-sided stone tower. Inside the tower were three wooden crucifixes suspended at different levels against the wall.

When I focused on the middle crucifix, an explosion of intense white light passed through me, creating a powerful vibration that I can only equate to an electric current. From the middle of the light appeared an angel, his cherubic face coming closer until he kissed me on the mouth. Then he placed his hands on either side of my face and lifted me out of my body and off into space.

For a moment I felt complete euphoria, a sense of weightlessness and selflessness that was incredibly alluring until I almost felt like I was dying. Struggling free of the angel, I fell back into my body and awoke.

My heart was pounding and sweat soaked the bed. I got up immediately. I was badly shaken, and every way I turned it, the only meaning I could deduce from the dream was that I was going to die. It was still dark outside, and I didn't want to wake Denholm. Unable to go back to sleep, I sat down at the desk and began writing a note to Margot and my children, telling them about the dream and my interpretation of it. By its end the letter had turned into a will.

Finally I fell into a fitful sleep, waking up as the sun removed the veil of darkness from the jungle. Looking out the door at the placid blue ocean, I pondered what I should do. *Should I dive today, or should I stay out of the water?*

DE BRY HAD ARRIVED on an early-morning plane in time for breakfast. As we ate, I told him and Denholm about my dream. I even showed them the notebook in which I had written the note, with handwriting all over the page, like the writing of a madman.

They listened patiently as I described the crosses and the angel and leaving my body. They acknowledged that the events of the last week had been intense and could certainly lead to a variety of bizarre and vivid dreams. And then there was always the Lariam. Still, de Bry said, he didn't want to ignore the possibility that there was some kind of meaning in the dream. "Maybe it doesn't mean you're going to die," he said. "Maybe it means something entirely different."

We packed our gear, piled into the Land Rover, and headed for the wreck site. As we bumped over the dirt road to the bay and passed Swete's bungalow, I couldn't help but remember how quickly circumstances change. A few days earlier I had been filled with anger for Swete. Now those emotions had turned to pity and empathy as I remembered him melting away in that hellhole of a hospital.

At the wreck site we joined Maximo Felice, who anchored his dive boat between the two wrecks. Felice and Denholm stayed with the boat to watch for any potential danger from the surface, while de Bry and I dove on the site.

He chose the easternmost wreck because he thought it was most likely the *Fiery Dragon*. "A wreck with coins is always worth diving on," he said. Coins mean dates and dates mean identification.

I was fine with the wreck site to the west. Just as Swete had come to this island for the *Serapis,* I had come for the *Adventure Galley.* If this was the *Adventure Galley,* it would offer little if anything of value, since the pirates on both of Kidd's ships had stripped them of their valuables. It may sound odd, but it was more important to me to dive on the remains of Kidd's ship than it was to collect artifacts.

We turned on our air and flipped off the boat. De Bry went his way and I went mine. Skimming across the bottom, we each found our respective pile of ballast stones and began digging test pits, holes that would reveal any artifacts that might be buried in the ballast mound.

I dug in the mound without finding much until my tank was nearly out of air. Then I headed back to the dive boat to get a fresh supply of air. I was pulled on board by Felice and Denholm. A few minutes later de Bry came up, his net bag containing an odd-looking artifact.

"Hey, Barry, you're not going to believe this," he said as he, too, was pulled into the boat. Sitting down next to me, he opened his net bag and took out a wooden figure of Jesus Christ. It was about eighteen inches long and had broken from the crucifix, but otherwise it was in good condition for having been underwater for nearly three hundred years.

"So this is what that dream meant," I said, relief flooding through me. "The first artifact found would be a crucifix."

Finding the artifact of my dream was a good omen. We quickly donned fresh air tanks and went back to our respective ballast mounds. An hour later we surfaced again for fresh tanks. Once again de Bry had that impish look on his face. Taking off his gloves, he poured out a handful of gold coins. It was exactly the kind of mix that one would find

BARRY CLIFFORD DISPLAYS A GOLD COIN FOUND ON THE *FIERY DRAGON* SITE.
Nick Caloyianis

on the wreck site of a ship that had robbed vessels from around the world. There were currencies from Germany, Italy, the Netherlands, Austria, Arabia, India, the Ottoman Empire, and other places.

I could imagine the crowded and messy living conditions that existed below decks. Crew members sometimes carried their coins in sock rolls, and it wasn't difficult to imagine some of those makeshift purses being overlooked as the pirates gathered their booty for the transfer to shore.

That night at the hotel we examined the coins more closely. Several of them had dates that could not be clearly deciphered. But one, a Dutch coin, showed a date of 1718, which put this site at nearly two decades after the sinking of the *Adventure Galley*. Given this new piece of information and the location of the wreck site, de Bry was certain that his wreck was the *Fiery Dragon*.

Just to make sure, he dug a deep test pit at the site the next day. During our second expedition in June 2000, he had noticed construc-

THE ARTIFACT OF BARRY CLIFFORD'S DREAM, A WOODEN JESUS FOUND ON THE *FIERY DRAGON* SITE.   *John de Bry*

tion techniques on the hull that hinted at Dutch rather than English architecture. To confirm his suspicions, he exposed more of the ship's ribs until he could closely examine one of the futtocks, the curved timbers that form the frame of the hull. The way it was attached to the floor timber convinced him that this was a Dutch ship.

We now had enough evidence to make a positive identification. The first wreck site we had dove on, the one to the east, was the *Fiery Dragon*. That meant the one to the west had to be the *Adventure Galley*.

It was official. I had finally found Captain Kidd's flagship.

"How do you feel now, Barry?" asked de Bry. "You've found what you came for."

For a moment it didn't sink in. I was at a loss for words. Was there something profound, a quote from Captain Kidd perhaps, or from one of his backers? But none of them seemed to have ever focused on the profound, dedicating their lives instead to chasing money, status, and power.

Captain Kidd had everything he needed in New York—a wife, children, and enough property and income to support him for the rest of his life. But it wasn't enough. He threw away both his honor and his life trying to get more. I couldn't blame Kidd, not really. Driven like an old hunting dog, he had a lust for wealth that seemed almost instinctual. The same was true of his backers. They were all like old hunting dogs, nosing out as much power as they could.

In a way it was this that set Kidd apart from other pirate captains. They were elected by their men and therefore worked *for* their men. If they didn't, they were quickly voted out and replaced with someone who could better serve them. Kidd was not elected as a pirate captain; indeed, he didn't start out to be a pirate. He was never chosen by his men and never appeared to truly care about them. Instead, Kidd had sought appointment by the rich and powerful, in a bad enterprise, so that he himself could become richer and more powerful.

Kidd tried to have it both ways. He wanted riches and social standing regardless of the cost, committing acts of piracy and other crimes at the same time that he was trying to pass himself off to his backers as an honest privateer. In the end Kidd fooled only himself. Most of his men left him right here in Pirate Bay on Île Sainte-Marie to pursue a life of true piracy with a real pirate captain. His backers abandoned him, leaving him dangling at the end of a rope.

I couldn't find anything profound to say because the wreck site of the *Adventure Galley* was more than anything else a symbol of bad leadership, flawed planning, and corrupt motives. A sunken monument to greed. As the old saying goes: "'Tis skill, not strength, that governs a ship." Kidd never truly understood that, and in the end his failure to understand proved fatal.

It struck me that the two ships we found could not have been commanded by more different men. Captain William Condon of the *Fiery Dragon* was a "pirate's pirate," a man who remembered he had been chosen by his shipmates and deferred to them. Kidd was appointed by a group of investors and considered his crew employees, or less. Condon was able to negotiate well among the different factions of his crew as well as outsiders. Kidd was ill-tempered, an autocrat who tried to rule by blustering force. Condon was a risk taker and clearly not afraid to engage in combat. Kidd was fearful and indecisive. He was hanged in disgrace for piracy and murder, set up by the very people who hired him to ravage the seas. Condon retired from piracy and settled on the Normandy coast in France where he became a thriving merchant and shipowner, known locally for his "honor and probity."

I couldn't help but recall Defoe's report on the captain of the *Whydah,* Sam Bellamy, and his commentary on "the high and mighty": "They rob the poor under the cover of law, forsooth, and we plunder the rich under the protection of our own courage. Had you not better make one of us, than to sneak after the a—s of those villains for employment."

It is probably only coincidence that Bellamy's and Condon's biographies are side by side in the pirate masterpiece *A General History of the Pirates.*

Two different approaches to piracy and two different results, I thought. It seemed almost symbolic that Kidd's ship was practically devoid of artifacts, while Condon's seemed to be a rich trove that would someday reveal much about the ship and its era.

But all I told de Bry was that I was elated to confirm the find of the *Adventure Galley.* And by stumbling upon the *Fiery Dragon,* we had found twice as much history as we were looking for. *The search for shipwrecks is full of happy accidents,* I told myself. This had certainly been one of those.

WITHIN A FEW DAYS Paine arrived, as did the Discovery Channel camera crew. I was glad to see them. Now we could finally get the television footage we had come for. And with Paine we could increase the number of underwater hands, which were greatly needed. De Bry's malaria was making a comeback, giving him the chills. Even though he insisted on working right through the symptoms, it was clear that he was pushing his physical limits.

We got Paine into the water as soon as he arrived. Since we had confirmed the identity of both ships, we now wanted to focus on finding as many diagnostic artifacts as we could. Our ten-day permit period was nearing its end, and we felt that it would be more productive to gather artifacts than to dig test pits. We could examine artifacts later at leisure on the surface, but test pits could only be examined by actually diving on the site. Given our permit hassles, we weren't sure when that would happen again once the ten days were over, and we didn't want to expose any more material than could be promptly conserved.

By the end of our ten days, de Bry had recorded more than one hundred artifacts and artifact assemblages, meaning several related items labeled with a single tag. In addition to the wooden Christ figure, we found a complete blue-and-white Chinese export porcelain cup, a doughnut-shaped ceramic flask, a scale, a ceramic oil lamp, the bottom of an English gin bottle, and large piles of Chinese ceramic shards.

At the end of one dive, de Bry found a clay lion that was approximately eight centimeters long. It was purple and ornate and only slightly broken, a marvel for having spent 279 years in a sometimes rough undersea environment. Later, de Bry was able to use the lion's distinct purple color to tell that it came from Yixing, China, 120 miles northwest of Shanghai. Purple clay is still mined there today for statuary and tea sets. The porcelain proved to be blue-and-white cup fragments from the Kangxi period (1662–1722). Experts in France suspect that the ceramic cups came from the kilns of Ching-te Chen in southern China. This region is rich in deposits of fine clay and produced imperial ware under court patronage for centuries.

By analyzing all of the artifacts, de Bry was able to tell that much of

JEFF DENHOLM WRESTLES WITH A CABLE USED TO PULL UP THE MORE THAN
ONE HUNDRED ARTIFACTS FOUND, INCLUDING STATUES, PORCELAIN, AND
COINS FROM VARIOUS COUNTRIES. *John de Bry/Barry Clifford*

the cargo on the *Fiery Dragon* had been stolen in 1720 from the *Maison d'Autriche,* the four-hundred-ton, twenty-four-gun Hapsburg cargo ship that was on her way to Oostende from Canton, China, when she was taken in February.

Few artifacts were found on the *Adventure Galley,* with the exception of fragments of Chinese and European porcelain. These were slightly different in style from the ones found on the *Fiery Dragon* and suggested to de Bry that they were from an earlier dating.

On the next-to-last day of diving, though, Paine made a thrilling discovery on the *Adventure Galley* site. Moving some of the larger ballast rocks, he found an enormous pewter tankard. It was the kind of high-status object that likely belonged to an officer or maybe even Kidd himself. Heavily encrusted with marine life, it looked like a piece of modern art. Paine held it up so de Bry, who was on the boat, could see it.

"My God, it's Kidd's beer mug!" shouted de Bry when he saw the object.

AT THE END OF our tenth day of excavation we were exhausted. We went to our bungalows to shower and change our clothes, and then we all met in the dining room. None of us looked good. De Bry's case of chills had become serious. Denholm was sunburned from days on the boat and worn out from diving. Paine was tired from jet lag and the exertion of two days of constant diving. But we were happy. We had finally confirmed the site of the *Adventure Galley* and, as a bonus, the *Fiery Dragon.* Ultimately, the *Fiery Dragon* wreck will prove to be more fruitful from an archaeological standpoint, since the pirates of that vessel robbed far more ships than Kidd's crew ever dreamed of.

I knew that someday we would return to do a full excavation of this site, but for now I was happy to have identified both ships and retrieved what artifacts we had found.

I was sitting in the dining room with my back to the kitchen when

A TANKARD FOUND ON THE *ADVENTURE GALLEY* SITE.
Nick Caloyianis

Denholm, sitting across from me, suddenly looked up and shouted, "I can't believe what I'm seeing!" I looked at the faces of my companions. Paine began to laugh and applaud. De Bry had a look of pleasure and horror on his face at the same time.

I turned around to see Fifou approaching with Captain Kidd's tankard in one hand and an open bottle of rum in the other. He had a broad smile on his face as he approached the table and held the giant mug high. Standing next to us, he poured about half the bottle of rum into the tankard.

"You can't do that with an artifact!" declared de Bry, reaching up with both hands.

Paine got to it first. He held the tankard between his palms and took a hearty drink. When he was finished he handed it to Denholm to take a swig. I looked into the mug and paused for a moment. Swirling around in the rum was ocean sediment that Fifou had been unable to wash out. I put the tankard to my lips, gulping down far more than my share. I held it out to de Bry, but he refused to take it. "That's an important artifact," he said seriously. "We shouldn't drink out of it."

"Come on, John," said Paine. "Take a drink to the pirates."

John lightened up. "I will if it's to the pirates," he said. "To the brotherhood of pirates!"

He drained the tankard, sediment and all, and set it down in the middle of the table. For a brief moment, Captain Kidd had joined us at the table.

ACKNOWLEDGMENTS

The search for historic shipwrecks is the product of teamwork. To that end, I wish to thank the project team for the *Quest for Captain Kidd* expeditions: Charlie Burnham, Brandon Clifford, Jenny Clifford, Stephanie de Bry, Jeff Denholm, Maximo Felice and Dr. Claudia Allegra, Layne Hedrick, Chris Macort and Cathrine Harker Macort, Todd Murphy, Bob Paine, Ben Perry, Eric Scharmer, Wes Spiegel, and most of all Dr. John "Doc Turtle" de Bry for the skills and hard work he gave to the expeditions, as well as his attention to the archaeological and historical content of this book.

I also wish to thank the specialists who came with us to the Red Island: Al Witten, Jakob Haldorsen, Doug Miller, I. J. Won, and Warren Getler.

I also appreciate the efforts of all of the other individuals and companies who helped to make our discovery mission a success, including Benchmade Knives, the Divers' Alert Network, Gateway Computers, Leica, Les Lézards de Tana, Mares-Dacor, Maui Jim, Fifou Mayer of the Princesse Bora Lodge, Mountain Hard Wear, Olympus, Annick Ratsiraka, Spot Image, and Trimble.

A special debt is owed to Bob Rhule of the U.S. Embassy in Madagascar.

Other acknowledgments are required: David Allen, Andrew Buckley, Preston Burchard, Jean-Paul Desroches, and Cynthia Mayer Merlini—I am grateful for your help.

Though it makes no difference to him now, Theophilus Turner's willingness to talk to the authorities made all the difference to me.

Helping to tell this story were Dan Conaway of HarperCollins; David Conover and his team at Compass Light Productions; Abby Greensfelder, Steve Burns, Steve Manuel, Mike Quattrone, and all of the *Quest for Captain Kidd* production staff at Discovery Communications; my agent, Nat Sobel of Sobel-Weber Associates, and, most of all, Paul Perry of Paradise Valley, Arizona, who made all three trips to Africa with me.

Most especially, however, this project was made possible by my friend and projects historian Ken Kinkor, who first gave me the idea to search for the *Adventure Galley*—Ken's vivid description of where Kidd lost his ship conjured an image so real that the only cure for my fascination was to find the shipwreck.

BOOKS BY BARRY CLIFFORD

RETURN TO TREASURE ISLAND AND THE SEARCH FOR CAPTAIN KIDD
ISBN 0-06-095982-7 (paperback)

The fascinating, unbelievable story behind Barry Clifford's discovery of the long-lost treasure ship— the *Adventure Galley*—and of the world's most fabled pirate, Captain Kidd.

THE LOST FLEET
The Discovery of a Sunken Armada from the Golden Age of Piracy
ISBN 0-06-095779-4 (paperback)

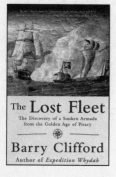

Alternating between the dramatic account of a maritime calamity in 1678—when the majority of the French fleet sank in the Caribbean—and Barry Clifford's own expedition to document the wrecks, *The Lost Fleet* brings to life a period in maritime history that had a profound effect on the shape of America.

EXPEDITION WHYDAH
The Story of the World's First Excavation of a Pirate Treasure Ship and the Man Who Found Her
ISBN 0-06-092971-5 (paperback)

In 1717, a small flotilla under the command of Captain Samuel Bellamy lost a desperate battle with the elements—sending Bellamy, the *Whydah*, and priceless treasure to a watery grave. Clifford alternates between the story of his own lifelong search for the *Whydah* and the history of the former slave ship, pirates, and life on the high seas.

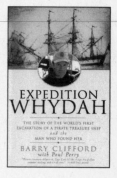